Springer Series in Statistics

Springer
New York
Berlin
Heidelberg
Barcelona
Hong Kong
London
Milan
Paris
Singapore
Tokyo

Springer Series in Statistics

(continued after index)

Martin A. Tanner

Tools for Statistical Inference

Methods for the Exploration of Posterior
Distributions and Likelihood Functions

Third Edition

With 47 Figures

Springer

Martin A. Tanner
Department of Statistics
Northwestern University
2006 Sheridan Road
Evanston, IL 60208 USA

Mathematics Subject Classification: 62F15, 62Fxx, 62Jxx

Library of Congress Cataloging-in-Publication Data
Tanner, Martin Abba, 1957–
    Tools for statistical inference: methods for the exploration of
posterior distributions and likelihood functions / Martin A. Tanner.
  — 3rd ed.
        p.   cm. — (Springer series in statistics)
    Includes bibliographical references and index.
    ISBN 0-387-94688-8 (hardcover: alk. paper)
    1. Bayesian statistical decision theory.   2. Mathematical
statistics.   I. Title.   II. Series.
    QA276.5.T36   1996
    519.5′42 – dc20   96-10784

Printed on acid-free paper.

Production managed by Hal Henglein, manufacturing coordinated by Jeffrey Taub.
Typeset by Bytheway Typesetting Services, Norwich, NY, using T$_E$X.
Printed and bound by Edwards Brothers, Ann Arbor, MI.
Printed in the United States of America.

9 8 7 6 5 4 3

ISBN 0-387-94688-8   Springer-Verlag   New York   Heidelberg   Berlin   SPIN 10762442

Preface

This book provides a unified introduction to a variety of computational algorithms for Bayesian and likelihood inference. In this third edition, I have attempted to expand the treatment of many of the techniques discussed. I have added some new examples, as well as included recent results. Exercises have been added at the end of each chapter.

Prerequisites for this book include an understanding of mathematical statistics at the level of Bickel and Doksum (1977), some understanding of the Bayesian approach as in Box and Tiao (1973), some exposure to statistical models as found in McCullagh and Nelder (1989), and for Section 6.6 some experience with conditional inference at the level of Cox and Snell (1989). I have chosen not to present proofs of convergence or rates of convergence for the Metropolis algorithm or the Gibbs sampler since these may require substantial background in Markov chain theory that is beyond the scope of this book. However, references to these proofs are given. There has been an explosion of papers in the area of Markov chain Monte Carlo in the past ten years. I have attempted to identify key references—though due to the volatility of the field some work may have been missed. Hopefully enough information has been presented to allow the reader to impute what is missing!

I would like to thank my colleagues Hendricks Brown, Brad Carlin, Lu Cui, Alan Gelfand, Jack Hall, Dan Heitjan, Ina Hoeschele, John Kolassa, Chris Ritter, Yodit Seifu, Bruce Tier, and Wing Wong for their comments/suggestions/input into this work. I extend thanks to Chip Heathcote for inviting me to the Australian National University, where I began work on this third edition. I also wish to thank the American Statistical Association, the Institute of Mathematical Statistics, Biometrika Trustees, the Biometric Society, and the Royal Statistical Society for permission to reprint material that appeared in the *Journal of the American Statistical Association, Statistical Science, Biometrika, Biometrics*, and *Applied Statistics*, respectively. This work was supported by NIH grant CA35464 and NSF grant DMS-9505799. I take responsibility for all errors and omissions.

בריך רחמנא דסייען

Evanston, Illinois
March, 1996

M. Tanner

Contents

1

Introduction

The goal of this book is to provide a unified introduction to a variety of computational algorithms that can be used as part of a Bayesian (posterior) analysis or as part of a likelihood analysis. These algorithms are tools and may be categorized using several taxonomies. The reader may find it useful to review these taxonomies as an aid to understanding how these tools complement, contrast and extend each other.

i. Function: Maximize, Marginalize and Simulate

The Newton–Raphson, *EM* (expectation maximization) and Monte Carlo *EM* (*MCEM*) algorithms are used to *locate the mode or modes* of the likelihood function or of the posterior density. Moreover, the Newton–Raphson algorithm as well as the *EM* and *MCEM* algorithms (with additional effort) provide information regarding the curvature of the likelihood at a mode. Other algorithms (e.g. Laplace's method, data augmentation and the Gibbs sampler) can be used to *obtain a marginal* of the likelihood function or of the posterior density. Alternatively, acceptance/rejection, data augmentation, the Gibbs sampler and the Metropolis algorithm can be used to *obtain a sample of parameter values* from the likelihood function or from the posterior density. This sample may then be used to obtain an estimate of a lower dimensional marginal of the likelihood or posterior, to estimate a moment of this marginal or, more generally, to estimate some functional of the likelihood or posterior (e.g. the content and boundary of the highest posterior density region or the predictive distribution).

ii. Random vs. Deterministic Algorithms

Alternatively, one can categorize these algorithms into non-Monte Carlo, noniterative Monte Carlo and iterative Monte Carlo methods. The non-Monte Carlo tools (e.g. Newton–Raphson, *EM* and Laplace) do not require the input of a stream of (pseudo) random numbers. The importance sampling and rejection/acceptance algorithms require a stream of (pseudo) random numbers as input and produce a sample from the likelihood function or from the posterior density as output. They are not iterative in nature. The Monte Carlo *EM* algorithm, data augmentation, the Gibbs sampler and the Metropolis algorithm all require a random input stream and require iteration to realize a sample from the likelihood function or from the posterior distribution of interest.

iii. Augmentation vs. Nonaugmentation Algorithms

In the *EM*, Monte Carlo *EM*, data augmentation and poor man's data augmentation algorithms, the data analyst augments the observed data with latent data to simplify the computations in the analysis. This latent data may be "missing" data (e.g. a missing data point in a two-way anova with one entry per cell), parameter values (e.g. the degrees of freedom of the *t* error distribution) or values of sufficient statistics. Readers of this book should understand that augmentation means more than filling in missing observations and that the augmentation vs. nonaugmentation taxonomy is not simply "missing data" vs. "observed data". Thus, in the *EM* algorithm, one augments the observed data with latent data such that one complicated maximization is replaced by an iterative series of simple maximizations. In the data augmentation algorithm, one augments the observed data with latent data to facilitate the sampling operation. The aim is to replace a complicated simulation by an iterative series of simple simulations.

Chapter 2 reviews the definitions of likelihood function and posterior density, the method of maximum likelihood, the Newton–Raphson algorithm for locating the mode of an objective function and provides a brief review of normal-based Frequentist and Bayesian inference. Chapter 3 motivates and illustrates Laplace's method as a technique for calculating higher-order approximations to the likelihood or to the posterior. This methodology allows for non-normal approximations to the likelihood or to the posterior density. The noniterative Monte Carlo, importance sampling and rejection/acceptance methods are presented as tools which yield a sample from the likelihood or from the posterior. Thus, the algorithms in Chapters 2 and 3 do not make use of any augmentation of the observed data to define the algorithm. These methods are applied directly to the likelihood or to the posterior density.

Chapter 4 presents the first of the augmentation algorithms—the *EM* algorithm. We motivate and illustrate this algorithm, and also present various methods for calculating standard errors in the context of the *EM* algorithm. In Chapter 5 the data augmentation algorithm is motivated and illustrated and issues relating to assessment of convergence are presented. Also presented are the poor man's data augmentation algorithms which reduce the computational burden of the data augmentation algorithm, at the expense of providing a sample from an approximation to the likelihood or to the posterior. Importance sampling alternatives (e.g. *SIR* and sequential imputation) are also presented.

Since the data augmentation algorithm (Chapter 5) can be viewed as a two-component version of the Gibbs sampler, we follow the data augmentation algorithm with a presentation of general Markov chain Monte Carlo methods: the Gibbs sampler and the Metropolis algorithm (Chapter 6). In fact, one may argue that the Gibbs sampler is an augmentation algorithm in that one augments the observed data with parameters (the latent data). In other words, rather than operating on the likelihood or the posterior $p(\theta_1, \theta_2, \ldots, \theta_d \mid Y)$, the Gibbs sampler requires that one augments the observed data Y with parameters to obtain, for example, $p(\theta_1 \mid \theta_2, \ldots, \theta_d, Y)$. Simply put, in augmentation one puts "stuff" on the right-hand side of the vertical bar. The Metropolis algorithm is not an augmentation algorithm, since it works directly with $p(\theta_1, \theta_2, \ldots, \theta_d \mid Y)$, modulo the normalizing constant. Both the Gibbs sampler and the Metropolis algorithm yield a Markov chain whose equilibrium distribution (under certain regularity conditions) is the normalized likelihood function or the posterior density of interest. Thus, both algorithms provide the data analyst with a correlated sample from the function of interest. Also discussed are techniques for assessing convergence and approaches for handling the situation where one cannot directly sample from the conditional distributions as required in the Gibbs sampler.

The Frequentist, in contrast to the Bayesians or to the likelihoodists (i.e. those who do base their inference on the shape of the likelihood), may have little interest in the bulk of this book. First, the Frequentist may be put off by the use of the posterior density in the presentation of the algorithms throughout this book. In this regard, it should be understood that while we show how the *EM* algorithm can be used to maximize the posterior, one can also use this algorithm to maximize the likelihood—though in any given situation the posterior calculation may be simpler due to the specification of the prior. Clearly, the mode of the likelihood and the curvature at the mode are of great value to the Frequentist's *large sample (normal-based)* inference approach. However, the Frequentist may have little interest in the marginalization uses of data augmentation, the Gibbs sampler and the Metropolis algorithm, since integrating out parameters of the likelihood may not be a sensible *Frequentist* calculation. Thus, while to the Bayesian or to the likelihoodist tools such as data augmentation, Gibbs sampler, and Metropolis are of immense value to the analysis of a small sample problem, they are of little direct help to the Frequentist since they do not determine the distribution of an estimator (e.g. the maximum likelihood estimator) or of a test statistic (e.g. likelihood ratio) *in repeated (small) samples of data*. With this concern in mind, the final section of this book illustrates how the Gibbs sampler, in conjunction with saddlepoint methods, may be used to simulate sufficient statistics within the conditional Frequentist inference paradigm. In this way, we illustrate the application of Markov chain algorithms to small sample Frequentist inference.

We now consider three examples which will help to motivate these various tools for statistical inference and which will be examined later in this book.

EXAMPLE. *Censored Regression Data*
The Stanford Heart Transplant Program began in October 1967. The patients in the program were admitted after review by a committee and they then waited for donor hearts to become available. Some patients died while waiting for a transplant, but most received a transplant. The data presented in Table 1.1 summarize survival time in days after transplant for 45 patients. (See Miller (1980) or Miller and

TABLE 1.1.   Stanford Heart Transplant Data, February 1980.

Patient no.	Survival time, days	Status*	Age at transplant	Mismatch score**
1	15	1	54	1.11
2	3	1	40	1.66
3	46	1	42	0.61
4	623	1	51	1.32
5	126	1	48	0.36
6	64	1	54	1.89
7	1350	1	54	0.87
8	23	1	56	2.05
9	279	1	49	1.12
10	1024	1	43	1.13
11	10	1	56	2.76
12	39	1	42	1.38
13	730	1	58	0.96
14	1961	1	33	1.06
15	136	1	52	1.62
16	1	1	54	0.47
17	836	1	44	1.58
18	60	1	64	0.69
19	3695	0	40	0.38
20	1996	1	49	0.91
21	0	1	41	0.87
22	47	1	62	0.87
23	54	1	49	2.09
24	51	1	50	—
25	2878	1	49	0.75
26	3410	0	45	0.98
27	44	1	36	0.0
28	994	1	48	0.81
29	51	1	47	1.38
30	1478	1	36	1.35
31	254	1	48	1.08
32	897	1	46	—
33	148	1	47	—
34	51	1	52	1.51
35	323	1	48	1.82
36	3021	0	38	0.98
37	66	1	49	0.66
38	2984	0	32	0.19
39	2723	1	32	1.93

(*continued*)

TABLE 1.1.   *Continued*

Patient no.	Survival time, days	Status*	Age at transplant	Mismatch score**
40	550	1	48	0.12
41	66	1	51	1.12
42	65	1	45	1.68
43	227	1	19	1.02
44	2805	0	48	1.20
45	25	1	53	1.68

Source: Miller and Halpern (1982).
 *Status: dead, 1; alive, 0.
** —denotes missing score.

Halpern (1982) for an extensive treatment of this data set.) Available information for each patient include: survival time, an indication of whether the patient is dead or alive as of February 1980, the age of the patient in years at the time of transplant and a mismatch score which attempts to quantify the dissimilarity between donor and recipient tissues with respect to HLA antigens.

Suppose we wish to regress log survival time on age. We notice that five of the individuals in our reduced version of the data set are still alive as of February 1980. In other words, the failure times for these subjects are *censored*. We do not have a survival time for these people—we only know that a given person survived beyond the recorded event time. Thus, we cannot directly use standard regression routines (e.g. REGRESS in MINITAB) to analyze these data. To analyze these data, we must resort to more sophisticated methods.

In this book, we will discuss two approaches to the analysis of such data. In the nonaugmentation approach, we construct the loglikelihood function for a given model for the observed data. We then maximize this function to estimate the parameters of the model. Unfortunately, the loglikelihood may be quite complicated. For example, in the case of censored regression data with normal errors, the loglikelihood will contain a sum of logarithms of normal survivor functions. In an alternative approach, a data augmentation approach, we augment the data with unobserved or *latent* data and compute the expected value of the loglikelihood for the augmented data. We then maximize this augmented loglikelihood function. This *EM* algorithm of augmenting the data and maximizing the expected value of the augmented loglikelihood is then iterated. In the case of censored regression data with normal errors, this data augmentation approach reduces to an iterated series of least-squares computations to estimate the regression coefficients. By augmenting the data, we replace the complicated analysis required in the observed data approach by a series of simple analyses.

EXAMPLE. *Latent Class Analysis*
The data in Table 1.2 represent responses of 3181 participants in the 1972, 1973 and 1974 General Social Survey. The responses are cross classified by year of study (3 levels) and a dichotomous response (yes/no) to each of three questions.
    All three questions begin:

"Please tell me whether or not you think it should be possible for a pregnant woman to obtain a legal abortion if"

Question *A*:
she is married and does not want any more children

TABLE 1.2.   Subjects in the 1972–1974 General Social Surveys, Cross-Classified by Year of Survey and Responses to Three Questions on Abortion Attitudes.

Year (D)	Response to A	Response to B	Response to C	Observed count
1972	Yes	Yes	Yes	334
	Yes	Yes	No	34
	Yes	No	Yes	12
	Yes	No	No	15
	No	Yes	Yes	53
	No	Yes	No	63
	No	No	Yes	43
	No	No	No	501
1973	Yes	Yes	Yes	428
	Yes	Yes	No	29
	Yes	No	Yes	13
	Yes	No	No	17
	No	Yes	Yes	42
	No	Yes	No	53
	No	No	Yes	31
	No	No	No	453
1974	Yes	Yes	Yes	413
	Yes	Yes	No	29
	Yes	No	Yes	16
	Yes	No	No	18
	No	Yes	Yes	60
	No	Yes	No	57
	No	No	Yes	37
	No	No	No	430

Source: Haberman (1979, p. 559).

Question *B*:
   the family has a very low income and cannot afford any more children
Question *C*:
   she is not married and does not want to marry the man.

The traditional latent class model (Goodman, 1974) supposes that the four manifest variables (responses to questions *A*, *B*, and *C* and year) are conditionally independent given a dichotomous unobserved (latent) variable (e.g. the respondents' true attitude toward abortion—pro/anti). That is, if the value of the dichotomous latent variable is known for a given participant, knowledge of the participant's response to a given question provides no further information regarding the responses to either of the other two questions. Note that in this example, there is no "missing" data, as was the case in the previous example. It is the *model* which characterizes the unobserved (latent) data.

In Chapter 5, we will analyze these data using the data augmentation algorithm. In this data augmentation approach, one simulates latent data (conditional on simulated values of the parameters of the model) and then simulates parameter values (conditional on the simulated latent data). This procedure is then iterated to realize a *sample* of parameter values from the exact likelihood. We then use this sample of parameter values as our basis of inference.

EXAMPLE. *Hierarchical Models*
The data in Table 1.3 represent the weights of 30 young rats measured weekly for five weeks. Note that there are no "missing" data in this data set—all rats were weighed according to schedule. However, we will still use augmentation methods to analyze these data. In Chapter 6, we will apply the two-stage hierarchical model to these data:

First Stage:

$$Y_{ij} \sim N(\alpha_i + \beta_i x_{ij}, \sigma^2)$$

Second Stage:

$$\begin{pmatrix} \alpha_i \\ \beta_i \end{pmatrix} \sim N \left\{ \begin{pmatrix} \alpha_c \\ \beta_c \end{pmatrix}, \Sigma \right\},$$

where $i = 1, \ldots, 30$, $j = 1, \ldots, 5$; x_{ij} is the age in days of the ith rat for measurement j; $x_{i1} = 8$, $x_{i2} = 15$, $x_{i3} = 22$, $x_{i4} = 29$, $x_{i5} = 36$, $i = 1, \ldots, 30$; and Y_{ij} is the weight of the ith rat at measurement j. This model implies that at the "first stage" the data Y_{ij} follow a normal distribution with mean $\alpha_i + \beta_i x_{ij}$ and variance σ^2. At the "second stage", the individual specific parameters α_i and β_i are assumed to be drawn from a bivariate normal distribution. Note that in contrast to the previous example, this model does not hypothesize any missing or latent data. Rather, in our analysis of the data, the parameters α_c, β_c, and Σ play the role of the "missing" data. In our analysis, we augment the observed data with values for these parameters and then simulate values for α_i, β_i, and σ. Given values for α_i and β_i,

TABLE 1.3.   Rat Population Growth Data.

Rat	Week 1	Week 2	Week 3	Week 4	Week 5
1	151	199	246	283	320
2	145	199	249	293	354
3	147	214	263	312	328
4	155	200	237	272	297
5	135	188	230	280	323
6	159	210	252	298	331
7	141	189	231	275	305
8	159	201	248	297	338
9	177	236	285	340	376
10	134	182	220	260	296
11	160	208	261	313	352
12	143	188	220	273	314
13	154	200	244	289	325
14	171	221	270	326	358
15	163	216	242	281	312
16	160	207	248	288	324
17	142	187	234	280	316
18	156	203	243	283	317
19	157	212	259	307	336
20	152	203	246	286	321
21	154	205	253	298	334
22	139	190	225	267	302
23	146	191	229	272	302
24	157	211	250	285	323
25	132	185	237	286	331
26	160	207	257	303	345
27	169	216	261	295	333
28	157	205	248	289	316
29	137	180	219	258	291
30	153	200	244	286	324

Source: Gelfand et al. (1990).

we simulate the "missing" data, i.e. α_c, β_c, and Σ. This procedure is then iterated to realize a sample of parameter values from the exact posterior distribution, which we will use for such tasks as approximating various marginals of the posterior.

Appendix

In this appendix, we list several distributions that will be of use throughout this book.

1. *Binomial* $\theta \sim \text{Bin}(n, p)$

$$p(\theta) = \binom{n}{\theta} p^\theta (1-p)^{n-\theta} \qquad E(\theta) = np, \quad \text{var}(\theta) = np(1-p)$$
$\theta = 0, 1, 2, \ldots, n$
$p \in [0, 1]$

2. *Beta* $\theta \sim \text{Beta}(\alpha, \beta)$

$$p(\theta) = \frac{\Gamma(\alpha + \beta)}{\Gamma(\alpha)\Gamma(\beta)} \theta^{\alpha-1}(1-\theta)^{\beta-1} \qquad E(\theta) = \frac{\alpha}{\alpha + \beta}$$

$\theta \in [0, 1],$

$\alpha > 0, \beta > 0$
$$\text{var}(\theta) = \frac{\alpha\beta}{(\alpha + \beta)^2(\alpha + \beta + 1)}$$

where,
$\Gamma(\alpha) = \int_0^\infty y^{\alpha-1}e^{-y}dy;$
$\Gamma(\alpha) = (\alpha - 1)\Gamma(\alpha - 1);$
$\Gamma(1) = 1; \Gamma\left(\frac{1}{2}\right) = \sqrt{\pi}$

Note: When $\alpha = \beta = 1$, the uniform distribution on $(0,1)$ is obtained.
Note: For future reference, let $Be(\alpha, \beta) = \frac{\Gamma(\alpha)\Gamma(\beta)}{\Gamma(\alpha+\beta)}$.

3. *Poisson* $\theta \sim \text{Poisson}(\lambda)$

$$p(\theta) = \exp(-\lambda)\frac{\lambda^\theta}{\theta!} \qquad E(\theta) = \lambda, \quad \text{var}(\theta) = \lambda$$
$\theta = 0, 1, 2, \ldots,$
$\lambda > 0$

4. *Gamma* $\theta \sim \text{Gamma}(\alpha, \beta)$

$$p(\theta) = \frac{\beta^\alpha}{\Gamma(\alpha)} \theta^{\alpha-1}e^{-\beta\theta} \qquad E(\theta) = \frac{\alpha}{\beta}, \quad \text{var}(\theta) = \frac{\alpha}{\beta^2}$$
$\alpha > 0, \beta > 0, \theta > 0$

Note: The gamma distribution $G(\alpha, \beta)$ is sometimes parameterized as:

$$p(\theta) = \frac{\beta^{-\alpha}}{\Gamma(\alpha)} \theta^{\alpha-1}e^{-\theta/\beta}$$

5. *Normal* $\theta \sim N(\mu, \sigma^2)$

$$p(\theta) = \frac{1}{\sqrt{2\pi}\sigma} \exp\left(-\frac{1}{2\sigma^2}(\theta - \mu)^2\right) \qquad E(\theta) = \mu, \quad \text{var}(\theta) = \sigma^2$$

$-\infty < \theta < +\infty,$
$-\infty < \mu < +\infty, \quad \sigma > 0$

6. *Inverse Gamma* $\theta \sim$ Inverse-gamma(α, β)

$$p(\theta) = \frac{\beta^\alpha}{\Gamma(\alpha)} \theta^{-(\alpha+1)} e^{-\beta/\theta} \qquad E(\theta) = \frac{\beta}{\alpha - 1}$$

$\alpha > 0, \beta > 0, \theta > 0$ $\qquad\qquad\qquad\qquad \text{var}(\theta) = \dfrac{\beta^2}{(\alpha - 1)^2(\alpha - 2)}$

Note: When $\alpha = \beta = 0$, the *improper* distribution proportional to $1/\theta$ is obtained.

7. *Chi-square* $\theta \sim \chi_\nu^2 \equiv$ Gamma$(\alpha = \nu/2, \beta = 1/2)$

$$p(\theta) = \frac{2^{-\nu/2}}{\Gamma(\nu/2)} \theta^{\nu/2-1} e^{-\theta/2} \qquad E(\theta) = \nu, \quad \text{var}(\theta) = 2\nu$$

$\nu > 0, \theta > 0$

8. *Inverse chi-square* $\theta \sim \chi_\nu^{-2} \equiv$ Inverse-gamma$(\alpha = \nu/2, \beta = 1/2)$

$$p(\theta) = \frac{2^{-\nu/2}}{\Gamma(\nu/2)} \theta^{-(\nu/2+1)} e^{-1/(2\theta)} \qquad E(\theta) = \frac{1}{\nu - 2},$$

$\theta > 0, \quad \nu > 0$ $\qquad\qquad\qquad\qquad\qquad \text{var}(\theta) = \dfrac{2}{(\nu - 2)^2(\nu - 4)}$

9. *Multivariate Normal* $\theta \sim N_k(\mu, \Sigma)$

$$p(\theta) = (2\pi)^{-k/2} |\Sigma|^{-1/2} \exp\left(-\frac{1}{2}(\theta - \mu)^T \Sigma^{-1}(\theta - \mu)\right) \qquad E(\theta) = \mu,$$

$-\infty < \theta_i < +\infty \quad i = 1, 2, \ldots, k, \ -\infty < \mu_i < +\infty, \qquad \text{var}(\theta) = \Sigma$
Σ is a symmetric, positive definite $k \times k$ matrix

Note: If $\mathbf{y} = (\mathbf{y}_1, \mathbf{y}_2)^T$, $\dim(\mathbf{y}_i) = d_i$, $d_1 + d_2 = d$, $\quad \mu = (\mu_1, \mu_2)^T$, and

$$\Sigma = \begin{pmatrix} \Sigma_{11} & \Sigma_{12} \\ \Sigma_{12}^T & \Sigma_{22} \end{pmatrix}, \quad \text{then}$$

9.1 $p(\mathbf{y}_1) = N_{d_1}(\mu_1, \Sigma_{11})$
9.2 $p(\mathbf{y}_1|\mathbf{y}_2) = N_{d_1}(\mu_1 + \Sigma_{12}\Sigma_{22}^{-1}(\mathbf{y}_2 - \mu_2), \Sigma_{11} - \Sigma_{12}\Sigma_{22}^{-1}\Sigma_{12}^T)$
9.3 $p(\mathbf{a}^T\mathbf{y}) = N_1(\mathbf{a}^T\mu, \mathbf{a}^T\Sigma\mathbf{a})$

Note: Let $\Sigma = AA^T$ (Cholesky decomposition of Σ). If $\mathbf{Z} = (Z_1 \cdots Z_d)^T$, where $Z_i \overset{\text{iid}}{\sim} N(0, 1)$, then $\theta = \mu + A\mathbf{Z}$ is $N_d(\mu, \Sigma)$.

10. *Multivariate t* $\theta \sim t_\nu(\mu, \Sigma)$

$$p(\theta) = \frac{\Gamma((\nu + k)/2)}{\Gamma(\nu/2)\nu^{k/2}\pi^{k/2}} |\Sigma|^{-1/2}\left(1 + \frac{1}{\nu}(\theta - \mu)^T \Sigma^{-1}(\theta - \mu)\right)^{-(\nu+k)/2}$$

$-\infty < \theta_i < +\infty \quad i = 1, \ldots, k,$ $\qquad E(\boldsymbol{\theta}) = \boldsymbol{\mu}, \quad \text{var}(\boldsymbol{\theta}) = \Sigma \dfrac{v}{v - 2}$

$-\infty < \mu_i < +\infty, v > 0$

Note: When $k = v = 1$, we obtain the Cauchy distribution.

Note: If $\mathbf{x} \sim N_k(\boldsymbol{\mu}, \Sigma)$, $y \sim \chi_v^2$ (independent of $\mathbf{x}$) and $u_i = x_i/(y/v)^{\frac{1}{2}}$ for $i = 1, \ldots, k$, then $\mathbf{u} \sim t_v(\boldsymbol{\mu}, \Sigma)$.

11. *Wishart* $\qquad W \sim \text{Wishart}_v(S) \qquad (\dim(W) = \dim(S) = k \times k)$

$$p(W) = 2^{-vk/2} \pi^{-k(k-1)/4} |S|^{-v/2} |W|^{(v-k-1)/2} \exp\left(-\frac{1}{2} \text{tr}(S^{-1} W)\right) \times$$

$$\left(\prod_{i=1}^{k} \Gamma((v + 1 - i)/2)\right)^{-1}$$

$\quad v > 0$, S and W symmetric positive definite $\qquad E(W) = vS$

Note: If $\boldsymbol{\alpha}_1, \ldots, \boldsymbol{\alpha}_v \overset{iid}{\sim} N_k(\mathbf{0}, S)$, then $\theta = \sum_{i=1}^{v} \boldsymbol{\alpha}_i \boldsymbol{\alpha}_i^T \sim W_v(S)$

Note: If $k = 1$, $S = 1$, then we obtain the χ_v^2 distribution.

12. *Inverse Wishart* $\qquad W \sim \text{Inverse-Wishart}_v(S^{-1})$

$$p(W) = 2^{-vk/2} \pi^{-k(k-1)/4} |S|^{v/2} \times$$

$$|W|^{-(v+k+1)/2} \exp\left(-\frac{1}{2} \text{tr}(SW^{-1})\right) / \left[\prod_{i=1}^{k} \Gamma((v + 1 - i)/2)\right]$$

$$E(W) = (v - k - 1)^{-1} S$$

13. *Multinomial* $\qquad \boldsymbol{\theta} \sim \text{Multinomial}(n, p_1, \ldots, p_k)$

$p(\boldsymbol{\theta}) = \begin{pmatrix} n \\ \theta_1 \theta_2 \cdots \theta_k \end{pmatrix} p_1^{\theta_1} \cdots p_k^{\theta_k}$ $\qquad E(\theta_j) = np_j,$

$\theta_1 + \cdots + \theta_n = n,$ $\qquad\qquad\qquad \text{var}(\theta_j) = np_j(1 - p_j)$

$p_1 + \cdots + p_k = 1, 0 \le p_i \le 1$ $\qquad \text{cov}(\theta_i, \theta_j) = -np_i p_j$

14. *Dirichlet* $\qquad \boldsymbol{\theta} \sim \text{Dirichlet}(\alpha_1, \ldots, \alpha_k)$

$p(\boldsymbol{\theta}) = \dfrac{\Gamma(\alpha_1 + \cdots + \alpha_k)}{\Gamma(\alpha_1) \ldots \Gamma(\alpha_k)} \theta_1^{\alpha_1 - 1} \cdots \theta_k^{\alpha_k - 1}$ $\quad E(\theta_j) = \dfrac{\alpha_j}{\alpha_0},$

$0 < \theta_i < 1, \Sigma_{j=1}^{k} \theta_j = 1, \alpha_j > 0,$ $\qquad \text{var}(\alpha_j) = \dfrac{\alpha_j(\alpha_0 - \alpha_j)}{\alpha_0^2(\alpha_0 + 1)}$

$\Sigma_{j=1}^{k} \alpha_j = \alpha_0$ $\qquad\qquad\qquad \text{cov}(\alpha_i, \alpha_j) = -\dfrac{\alpha_i \alpha_j}{\alpha_0^2(\alpha_0 + 1)}$

Exercises

Many of the exercises in this book will require you to generate (pseudo) random deviates. There are many tests to examine the adequacy of a random number generator. Exercises 1 and 2 are two such tests.

1. Generate 10,000 deviates from the U[0,1] distribution and plot x_i versus x_{i-1}. Is there any indication of dependence between successive values?

2. a. Simulate 10,000 random walks of length 50 in R^2 starting at (0,0). Compute the proportion of walks that end up in each of the four quadrants. What values would you expect for these proportions? Do the observed proportions vary significantly from the expected values? (See Vattulainen, Ala-Nissila and Kankaala, 1994.)
   b. Repeat (a) using walks of length 500.
   c. Repeat (a) and (b) using random walks on the lattice (i.e. restrict the jumps to unit increments) in R^2.

3. From each of the following distributions: (i) draw 10,000 deviates; (ii) compute the sample average and sample SD and compare these numbers to the true values; (iii) draw a histogram of the deviates along with a plot of the probability density function for the given distribution:

   binomial(14, .9); beta(.5, .5); gamma(12, 2); IG(12, 2); χ^2 on 2 df;

   $$\chi^{-2}\text{on 2 df.}$$

4. a. Draw a sample of size 10,000 from the trivariate normal distribution with mean (0,0,0) and variance-covariance matrix:

   $$\begin{pmatrix} 1 & 4.5 & 9 \\ 4.5 & 25 & 49 \\ 9 & 49 & 100 \end{pmatrix}$$

   b. Draw a histogram of the X_1 deviates, along with the true marginal. Compute the sample average and sample SD and compare these numbers to the true values.
   c. Draw a sample of size 10,000 from the conditional distribution $p(X_1 \mid X_2, X_3)$, take $X_2 = X_3 = 1$. Compute the sample mean and sample SD and compare these numbers to the true values. Draw a histogram of the simulated values, along with the true conditional distribution.
   d. Use the sample of size 10,000 from the joint trivariate distribution to obtain the conditional mean and SD of X_1 given $X_2 = X_3 = 1$, i.e. look at all the triples whose second and third components are within an ε of 1. How do these values compare to the true mean and SD for various values of ε?

5. Suppose Y has a Cauchy distribution with density

   $$f(y) = \frac{1}{\pi(1 + y^2)}$$

   (see Ripley, 1987). We wish to estimate $\theta = P(Y > 2) = 0.1476$. We consider four estimates of θ, where each is of the form

   $$\frac{1}{m} \sum_{i=1}^{m} \phi(x_i),$$

   where x_i is drawn from some distribution. Estimate θ via the following three approaches and compare the estimates using the sample standard deviation of the $\phi(.)$ values. Take $m = 10,000$.

a. $\phi(X) = 1$ if $X > 2$ and zero otherwise, where X is Cauchy. Show that the expectation of $\phi(X)$ is θ.

b. $\phi(X) = \frac{1}{2}$ if $|X| > 2$ and zero otherwise, where X is Cauchy. Show that the expectation of $\phi(X)$ is θ.

c. $\phi(X) = \frac{1}{2} f(X)$, where X is U[0, $\frac{1}{2}$]. Show that the expectation of $\phi(X)$ is θ.

2

Normal Approximations to Likelihoods and to Posteriors

In this chapter, we review several elementary concepts and methods from mathematical statistics. In Section 2.1, the likelihood and loglikelihood functions are defined and illustrated. The definition and illustration of the posterior density are also presented. Noninformative and conjugate priors are discussed in Section 2.2. In Section 2.3, the method of maximum likelihood is defined and the Newton–Raphson algorithm is presented as an algorithm for locating the posterior/likelihood mode. Section 2.4 presents Frequentist and Bayesian justification for using the normal approximation to the likelihood or to the posterior as the basis for inference. The delta method is defined and illustrated in Section 2.5 and Section 2.6 reviews the notion of the Highest Posterior Density region, a Bayesian approach to confidence regions and significance levels.

2.1 Likelihood/Posterior Density

Let Y denote the data, which can be scalar, vector valued or matrix valued and suppose that

$$Y \sim f(Y \mid \theta),$$

where $f(\cdot \mid \cdot)$ is a density function indexed by a parameter θ (scalar or vector).

Definition 2.1.1. Given the data Y, the likelihood function $L(\theta \mid Y)$ is any function of θ proportional to $f(Y \mid \theta)$.

The likelihood is a set of functions which differ by a factor that does not depend on θ. In other words, the likelihood function is defined up to a multiplicative constant. It is only the relative value (shape) of the likelihood which is of importance.

Definition 2.1.2. The loglikelihood function $l(\theta \mid Y)$ is the natural logarithm of $L(\theta \mid Y)$.

Definition 2.1.3. The normalized likelihood is equal to

$$\frac{L(\theta \mid Y)}{\int L(\theta \mid Y)d\theta},$$

when $\int L(\theta \mid Y)d\theta$ is finite.

EXAMPLE. *Univariate Sample from a Normal Population*
The joint density of n independent and identically distributed observations from a normal population with mean μ and variance σ^2 is given by

$$f(Y \mid \mu, \sigma^2) = (2\pi\sigma^2)^{-n/2} \exp\left[-\frac{1}{2}\sum_{i=1}^{n}\left(\frac{y_i - \mu}{\sigma}\right)^2\right].$$

For a given data set Y, the loglikelihood is

$$l(\mu, \sigma^2 \mid Y) = \frac{-n}{2}\log\sigma^2 - \frac{1}{2}\sum_{i=1}^{n}\frac{(y_i - \mu)^2}{\sigma^2}.$$

The loglikelihood is a function of μ, σ^2 for fixed Y.

EXAMPLE. *Randomly Censored Data*
Let $T_1, T_2, \ldots, T_n$ be *iid* with density f, cdf F and survivor function $S = 1 - F$. Thus, $S(t)$ represents the probability of surviving beyond time t. Let $C_1, \ldots, C_n$ be *iid* with density g and cdf G. T_i is the survival time for the ith individual and C_i is the censoring time associated with and independent of T_i. Subjects in a clinical trial may be censored due to loss to follow up (i.e. they move elsewhere), drop out (due to adverse effects the treatment is discontinued), or termination of the study. We observe $(Y_1, \delta_1), \ldots, (Y_n, \delta_n)$, where

$$Y_i = \min(T_i, C_i)$$

and

$$\delta_i = \begin{cases} 1 & T_i \leq C_i \\ 0 & T_i > C_i \end{cases}.$$

The pair (y_i, δ_i) has likelihood

$$f(y_i)^{\delta_i} S(y_i)^{1-\delta_i}[1 - G(y_i)]^{\delta_i} g(y_i)^{1-\delta_i}$$

and the likelihood of the entire sample is

$$\prod_u f(y_i) \prod_c S(y_i) \prod_c g(y_i) \prod_u [1 - G(y_i)],$$

where "u" denotes the uncensored observations and "c" denotes the censored observations. When the censoring distribution does not involve the unknown lifetime parameters, the last two products can be treated as constants. This is a reasonable assumption in a situation where patients randomly enter a study and a decision is made to terminate the study at a later date independent of the results of the study.

The assumption may not be reasonable if patients drop out of the study due to the toxicity of the therapy.

We now define the posterior density. In the Bayesian approach one conditions on the observed data and takes the parameters as random variables. Suppose that θ has a probability distribution $p(\theta)$. Then we have that

$$p(Y \mid \theta)p(\theta) = p(Y, \theta) = p(\theta \mid Y)p(Y).$$

Conditional on the observed data Y, the distribution of θ is

$$p(\theta \mid Y) = \frac{p(Y \mid \theta)p(\theta)}{p(Y)}. \tag{2.1.1}$$

Expression (2.1.1) is known as Bayes' theorem. Given the data Y, $p(Y \mid \theta)$ may be regarded as a function of θ, which we recall is the likelihood function of θ given Y and write as $L(\theta \mid Y)$. In this way, we have:

Definition 2.1.4. Given a likelihood $L(\theta \mid Y)$ and a prior probability density $p(\theta)$, the posterior density for θ is given as

$$p(\theta \mid Y) = cp(\theta)L(\theta \mid Y),$$

where

$$c^{-1} = \int_{\Theta} p(\theta)L(\theta \mid Y)d\theta,$$

when θ is continuous. When θ is discrete, we have $c^{-1} = \sum_{\theta} p(\theta)L(\theta \mid Y)$.

The quantity $p(\theta)$, which describes what is known about θ without knowledge of the data, is called the *prior distribution of* θ. The quantity $p(\theta \mid Y)$, which describes what is known about θ given the data Y, is called the *posterior density of* θ. Note that the *shape* of the posterior distribution is determined by the product of the likelihood and the prior.

EXAMPLE. *Normal Population (Mean Unknown, Variance Known)*
In this example, we consider the situation of an *iid* sample from a $N(\theta, \sigma^2)$ population, where σ^2 is known and θ is unknown. Note that $f(Y \mid \theta)$ is equal to

$$\prod_{i=1}^{n} \frac{1}{\sqrt{2\pi\sigma^2}} \exp\left\{-\frac{1}{2}\frac{(y_i - \theta)^2}{\sigma^2}\right\} = \frac{1}{(2\pi\sigma^2)^{n/2}} \exp\left\{-\frac{1}{2}\sum_{i=1}^{n}\frac{(y_i - \theta)^2}{\sigma^2}\right\}.$$

Since σ^2 is known, $(2\pi\sigma^2)^{-n/2}$ does not help to determine the shape of the likelihood. Hence we omit this term from our calculations. Also note that

$$\sum_{i=1}^{n}(y_i - \theta)^2 = \sum_{i=1}^{n}[(y_i - \bar{y}) + (\bar{y} - \theta)]^2 = \sum_{i=1}^{n}(y_i - \bar{y})^2 + n(\bar{y} - \theta)^2,$$

since $\sum_{i=1}^{n}(y_i - \bar{y})(\bar{y} - \theta) = \bar{y}\sum_{i=1}^{n} y_i - n\bar{y}^2 - \theta \sum_{i=1}^{n} y_i + n\theta\bar{y} = 0$. Thus, the likelihood can be reexpressed as

$$\exp\left\{-\frac{1}{2} \frac{\sum_{i=1}^{n}(y_i - \bar{y})^2 + n(\bar{y} - \theta)^2}{\sigma^2}\right\}.$$

Note that

$$-\frac{1}{2}\sum_{i=1}^{n} \frac{(y_i - \bar{y})^2}{\sigma^2}$$

is a constant (since as Bayesians we condition on the data). Hence, we have the further reduction that the likelihood can be expressed as

$$\exp\left\{-\frac{1}{2}\frac{n(\bar{y} - \theta)^2}{\sigma^2}\right\}.$$

Under the prior $p(\theta) \propto$ *constant*, it follows that the posterior density is $N(\bar{y}, \frac{\sigma^2}{n})$. Under the prior $p(\theta) = N(\theta_o, \sigma_o^2)$, where θ_o and σ_o are known constants, it can be shown (see Exercise 1) that the posterior density is normal with mean given by a weighted average of the prior mean and the sample average:

$$\theta_o \left(\frac{1/\sigma_o^2}{1/\sigma_o^2 + n/\sigma^2}\right) + \bar{y}\left(\frac{n/\sigma^2}{1/\sigma_o^2 + n/\sigma^2}\right),$$

and the posterior variance is equal to $(1/\sigma_o^2 + n/\sigma^2)^{-1}$. Note that as $\sigma_o \to \infty$ (n fixed), the prior flattens out and we recover the posterior under the prior $p(\theta) \propto$ constant. Similarly, as $n \to \infty$ (σ_o fixed) the "data swamp the prior".

As an illustration, Lee (1992) considers chest measurement data of 10,000 men. Based on his experience, Lee adopts a normal prior with mean of 38 inches and a standard deviation of 3 inches. The sample average of the data is 39.8 inches and the sample standard deviation is 2 inches, based on 10,000 observations. It is easy to check that the posterior mean is 39.8 inches, with a posterior standard deviation of $2/\sqrt{10,000}$. In this case, the "data swamp the prior," i.e. the posterior distribution is very close to the normalized likelihood.

EXAMPLE. *Linear Model*
The linear model is given as

$$y = x\theta + \varepsilon$$

where y is an $n \times 1$ vector of observations, θ is $d \times 1$ vector of unknown regression coefficients, x is an $n \times d$ matrix of known constants, ε is an $n \times 1$ vector of independent errors, where $\varepsilon_i \sim N(0, \sigma^2)$ and σ^2 is unknown. In this case,

$$f(Y \mid \theta, \sigma^2) = \left(\frac{1}{\sqrt{2\pi}}\right)^n \sigma^{-n} \exp\left[\frac{-1}{2\sigma^2}(y - x\theta)^T(y - x\theta)\right]$$

$$= \left(\frac{1}{\sqrt{2\pi}}\right)^n \sigma^{-n} \exp\left\{\frac{-1}{2\sigma^2}[vs^2 + (\theta - \hat{\theta})^T x^T x(\theta - \hat{\theta})]\right\}, \quad (*)$$

where

$$\hat{\theta} = (x^T x)^{-1} x^T y, \quad v = n - d$$
$$s^2 = (y - \hat{y})^T (y - \hat{y})/v \text{ and } \hat{y} = x\hat{\theta}.$$

Note that $\hat{\theta}$ is the least-squares estimate of θ, s^2 is the mean square error due to regression, and $\hat{y}$ is the predicted value of y at x. The reexpression of $f(Y \mid \theta, \sigma^2)$ in (*) follows since

$$(y - x\theta)^T (y - x\theta) = [(y - x\hat{\theta}) + (x\hat{\theta} - x\theta)]^T [(y - x\hat{\theta}) + (x\hat{\theta} - x\theta)]$$
$$= (y - x\hat{\theta})^T (y - x\hat{\theta}) + (\theta - \hat{\theta})^T x^T x (\theta - \hat{\theta}),$$

because $(\theta - \hat{\theta})^T x^T (y - x\hat{\theta}) = 0 = (y - x\hat{\theta})^T x (\theta - \hat{\theta})$.

We now present three major results relating to the posterior distribution for the linear model. These results will be used throughout the remainder of this book. The first result details the nature of the posterior distribution of θ and σ^2,—i.e. the posterior factors (neatly) into the product of an inverse chi-square distribution and a conditional normal distribution. The second result gives the marginal posterior distribution of θ having integrated out σ^2. The third result gives the posterior distribution of the quadratic form.

Result 2.1.1. Under the prior $p(\sigma^2, \theta) \propto \sigma^{-2}$, $p(\theta, \sigma^2 \mid Y) = p(\sigma^2 \mid s^2) \times p(\theta \mid \hat{\theta}, \sigma^2)$ where the marginal distribution of σ^2 is $vs^2 \chi_v^{-2}$ and the conditional distribution of θ, given σ^2, is $N(\hat{\theta}, \sigma^2(x^T x)^{-1})$.

Result 2.1.2. Under the prior $p(\sigma^2, \theta) \propto \sigma^{-2}$

$$p(\theta \mid Y) = \frac{\Gamma\left(\frac{v+d}{2}\right) \mid x^T x \mid^{1/2} s^{-d}}{\{\Gamma(1/2)\}^d \Gamma(v/2) v^{d/2}} \left[1 + \frac{(\theta - \hat{\theta})^T x^T x (\theta - \hat{\theta})}{vs^2} \right]^{-(v+d)/2}$$

i.e. the multivariate t distribution.

Result 2.1.3. Under the prior $p(\sigma^2, \theta) \propto \sigma^{-2}$, the posterior distribution of

$$\frac{(\theta - \hat{\theta})^T (x^T x)(\theta - \hat{\theta})}{ds^2}$$

follows the F distribution with d and v degrees of freedom.

To derive Result 2.1.1, note that $L(\theta, \sigma^2 \mid Y)p(\theta, \sigma^2) \propto$

$$\left[(\sigma^2)^{-d/2} \exp\left(-\frac{(\theta - \hat{\theta})^T x^T x (\theta - \hat{\theta})}{2\sigma^2} \right) \right]$$
$$\times [(\sigma^2)^{-(n-d)/2} \exp(-vs^2/2\sigma^2) \times (\sigma^2)^{-1}].$$

As a function of θ, the expression in the left brackets has the same shape as a multivariate normal density with mean $\hat{\theta}$ and variance–covariance matrix $\sigma^2(x^T x)^{-1}$.

As a function of σ^2, the expression in the right brackets has the same shape as the density of $(vs^2)/Z$, where Z is a χ_v^2 random variable. (Remember that as Bayesians we condition on the data so vs^2 is a constant.)

Result 2.1.1 has an interesting sampling interpretation. To draw a sample from the posterior density $p(\theta, \sigma^2 \mid Y)$, we draw z^* from the χ_v^2 and form $(vs^2)/z^* = \sigma_*^2$. We then draw θ^* from the (conditional) normal distribution $N(\hat{\theta}, \sigma_*^2(x^Tx)^{-1})$. The pair (θ^*, σ_*^2) is an observation from $p(\theta, \sigma^2 \mid Y)$.

Note that $p(\theta, \sigma^2 \mid Y)$ can always be factored as $p(\theta \mid \sigma^2, Y) \times p(\sigma^2 \mid Y)$. The contribution of Result 2.1.1 is that $p(\theta \mid \sigma^2, Y)$ and $p(\sigma^2 \mid Y)$ are standard/recognizable distributions and that the data Y are replaced by the sufficient statistics $(\hat{\theta}, s^2)$.

EXAMPLE. *Two-Sample Problem (Common Variance)*
We consider the situation where $x_1, \ldots, x_m$ is an *iid* sample from $N(\lambda, \sigma^2)$ and $y_1, \ldots, y_n$ is an independent *iid* sample from $N(\mu, \sigma^2)$, where λ, μ and σ are unknown. This situation is a special case of the linear model, since we can express the problem as

$$
\begin{bmatrix} x_1 \\ \cdot \\ \cdot \\ \cdot \\ x_m \\ y_1 \\ \cdot \\ \cdot \\ \cdot \\ y_n \end{bmatrix} = \begin{bmatrix} 1 & 0 \\ \cdot & \cdot \\ \cdot & \cdot \\ \cdot & \cdot \\ 1 & 0 \\ 0 & 1 \\ \cdot & \cdot \\ \cdot & \cdot \\ \cdot & \cdot \\ 0 & 1 \end{bmatrix} \begin{pmatrix} \lambda \\ \mu \end{pmatrix} + \begin{bmatrix} \varepsilon_1 \\ \cdot \\ \cdot \\ \cdot \\ \varepsilon_m \\ \varepsilon_{m+1} \\ \cdot \\ \cdot \\ \cdot \\ \varepsilon_{m+n} \end{bmatrix},
$$

where ε_i are *iid* $N(0, \sigma^2)$. Under the prior $p(\lambda, \mu, \sigma^2) \propto \frac{1}{\sigma^2}$, we have by Result 2.1.1 that $p(\lambda, \mu, \sigma^2 \mid data) = p(\sigma^2 \mid s^2) \times p(\lambda, \mu \mid \hat{\lambda}, \hat{\mu}, \sigma^2)$. It can be shown that $p(\lambda, \mu \mid \hat{\lambda}, \hat{\mu}, \sigma^2)$ is the bivariate normal distribution with mean $\binom{\bar{x}}{\bar{y}}$ and variance–covariance matrix

$$
\begin{pmatrix} \sigma^2/m & 0 \\ 0 & \sigma^2/n \end{pmatrix}
$$

and that $p(\sigma^2 \mid s^2)$ is the density of $\frac{(m+n-2)s^2}{\chi_{m+n-2}^2}$, where

$$
s^2 = \frac{\sum_{i=1}^{n}(y_i - \bar{y})^2 + \sum_{i=1}^{m}(x_i - \bar{x})^2}{m + n - 2}.
$$

If we let $\delta = \lambda - \mu$, then $p(\delta \mid \sigma^2, \bar{x}, \bar{y})$ is $N\left(\bar{x} - \bar{y}, \sigma^2\left(\frac{1}{m} + \frac{1}{n}\right)\right)$, by Result 9.3 of the Appendix in Chapter 1 (let $a = (+1, -1)$). It then follows by Result 2.1.2 that $p(\delta \mid data)$ is a t distribution on $m + n - 2$ df with mean $\bar{x} - \bar{y}$ and variance $s^2\left(\frac{1}{m} + \frac{1}{n}\right)$.

It is interesting to note that in this situation the Bayesian and Frequentist approaches yield similar solutions—a Frequentist would say that in *repeated samples* the pivotal

$$T = \frac{(\lambda - \mu) - (\bar{x} - \bar{y})}{s\sqrt{\left(\frac{1}{m} + \frac{1}{n}\right)}}$$

has a t distribution on $m + n - 2\,df$, independent of the values of λ, μ and σ. For the two-sample problem where the variances are not assumed to be equal, this parallel breaks down, since the distribution of the test statistic depends on the unknown ratio of the variances.

EXAMPLE. *Two-Sample Problem (Unequal Variances)*
We now consider the situation where $x_1, \ldots, x_m$ is an *iid* sample from $N(\lambda, \sigma^2)$ and $y_1, \ldots, y_n$ is an independent *iid* sample from $N(\mu, \rho^2)$, where λ, μ, σ and ρ are unknown. Under the priors $p(\lambda, \sigma^2) \propto \frac{1}{\sigma^2}$ and $p(\mu, \rho^2) \propto \frac{1}{\rho^2}$ we have that

$$p\left(\frac{\lambda - \bar{x}}{s_x/\sqrt{m}}\,\Big|\,data\right)$$

follows the t distribution on $m - 1\,df$ and that independently

$$p\left(\frac{\mu - \bar{y}}{s_y/\sqrt{n}}\,\Big|\,data\right)$$

follows the t distribution on $n - 1\,df$, where s_y is the standard deviation of the y values and s_x is the standard deviation of the x values.

It is straightforward to obtain the posterior distribution of $\lambda - \mu$ via simulation. In particular,

1. Draw t_* from the t distribution on $m - 1\,df$ and let $x = \bar{x} + \frac{s_x}{\sqrt{m}}t_*$.
2. Draw t_{**} from the t distribution on $n - 1\,df$ and let $y = \bar{y} + \frac{s_y}{\sqrt{n}}t_{**}$.
3. Let $z = x - y$.

Repeating steps 1–3 s times, yields an *iid* sample $z^{(1)}, \ldots z^{(s)}$ from $p(\lambda - \mu \mid data)$.

The analytic solution to this problem was obtained by Behrens (1929) and by Fisher (1935). Fisher justifies his solution via fiducial theory (see Zabell, 1992 for an extensive discussion). Let

$$T = \frac{(\lambda - \mu) - (\bar{x} - \bar{y})}{\sqrt{\frac{s_x^2}{m} + \frac{s_y^2}{n}}}$$

and let

$$\tan\theta = \frac{s_x/\sqrt{m}}{s_y/\sqrt{n}},$$

with θ in the first quadrant. It then follows (see Box and Tiao, 1973) that T has the Behrens–Fisher distribution ($BF(v_1, v_2, \phi)$) with parameters $v_1 = m - 1$, $v_2 = n - 1$ and $\phi = \theta$. The density of $BF(v_1, v_2, \phi)$ is

$$k \int_{-\infty}^{+\infty} \left[1 + \frac{(z \cos \phi - \tau \sin \phi)^2}{v_1}\right]^{-(v_1+1)/2} \times$$

$$\left[1 + \frac{(z \sin \phi + \tau \cos \phi)^2}{v_2}\right]^{-(v_2+1)/2} dz,$$

for $-\infty < \tau < +\infty$ and k^{-1} equal to $Be(v_1/2, 1/2) Be(v_2/2, 1/2)\sqrt{(v_1 v_2)}$. Lee (1992) presents tables for this distribution. See Scheffe (1970), Robinson (1976) and Wallace (1980) for a further discussion of the Behrens–Fisher problem.

Patil (1965) derived a t approximation to the Behrens–Fisher distribution by equating second and fourth moments. Let

$$f_1 = \left(\frac{v_2}{(v_2 - 2)}\right) \cos^2 \phi + \left(\frac{v_1}{(v_1 - 2)}\right) \sin^2 \phi,$$

$$f_2 = \left(\frac{v_2^2}{(v_2 - 2)^2(v_2 - 4)}\right) \cos^4 \phi + \left(\frac{v_1^2}{(v_1 - 2)^2(v_1 - 4)}\right) \sin^4 \phi,$$

$b = 4 + \frac{f_1^2}{f_2}$ and $a^2 = \frac{b-2}{b} f_1$. It then follows that $\frac{T}{a}$ is approximately a t distribution on $b\, df$.

From a Frequentist perspective, the distribution of

$$T = \frac{(\lambda - \mu) - (\bar{x} - \bar{y})}{\sqrt{\frac{s_x^2}{m} + \frac{s_y^2}{n}}}$$

in repeated sampling depends on the (unknown) ratio $\frac{\sigma}{\rho}$. The following approximation to this distribution due to Welch (1949) works well even for small m and n. (See Wang (1971) for a discussion of the adequacy of the approximation.) Approximately, T has a t distribution with $k\, df$, where k is equal to

$$\frac{\left(\frac{s_x^2}{m} + \frac{s_y^2}{n}\right)^2}{\frac{s_x^4}{m^2(m - 1)} + \frac{s_y^4}{n^2(n - 1)}}.$$

2.2 Specification of the Prior

The prior distribution represents an expression of belief about an unknown quantity before the data are available. In this section, we consider two classes of prior distributions: noninformative priors and conjugate (informative) priors.

2.2.1. Noninformative Priors

In situations where one does not have a strong prior belief or in public policy situations where prior opinions strongly differ, it is of use to have a neutral reference prior which is dominated by the likelihood.

Consider the case where $y_1, \ldots, y_n$ is an *iid* sample from $N(\mu, \sigma^2)$, where σ is known. As we saw in Section 2.1, the likelihood can be expressed as $\exp\left[\frac{-n}{2\sigma^2}(\bar{y} - \theta)^2\right]$. This is an example of a *data-translated likelihood*. When we plot this function (as a function of θ) we see that different data sets give rise to the same shape for the likelihood, except for a *shift* in location. In this case the data inform us (exclusively) as to the location of the likelihood. A noninformative prior (noninformative with respect to the information in the data) would therefore be ambiguous as to the location of the likelihood. A noninformative prior would suggest that any value is reasonable. Thus, in this case, it is reasonable to adopt the prior $p(\theta) \propto constant$.

In general, if we can write the likelihood $L(\theta \mid Y)$ in the form $g[\psi(\theta) - t(Y)]$, then the likelihood is in *data-translated form*, since different datasets only shift the likelihood through $t(Y)$. Arguing as above, we would take $p(\psi(\theta)) \propto constant$, since it is on the $\psi(\theta)$ scale that the likelihood is data-translated (i.e. on this scale the data inform us only where the likelihood is located)—thus, a noninformative prior would say that any value is reasonable *on the* $\psi(\theta)$ *scale*. Transforming back to the θ scale, we have $p(\theta) \propto |\frac{\partial \psi(\theta)}{\partial \theta}|$.

EXAMPLE. *Normal Population (Mean Known, Variance Unknown)*
Let $y_1, \ldots, y_n$ be an *iid* sample from $N(\mu, \sigma^2)$, then $L(\sigma \mid Y) = \sigma^{-n} \exp\left(\frac{-ns^2}{2\sigma^2}\right)$, where $s^2 = \sum_{i=1}^{n}(y_i - \theta)^2/n$. Now note that

$$\frac{\sigma^{-n}}{s^{-n}} \exp\left(\frac{-ns^2}{2\sigma^2}\right) = \exp\left[-n \log\left(\frac{\sigma}{s}\right)\right] \times \exp\left[\frac{-n}{2} \exp\left(\log\left(\frac{s^2}{\sigma^2}\right)\right)\right]$$

$$= \exp(-n(\log \sigma - \log s) - \frac{n}{2} \exp[-2(\log \sigma - \log s)]).$$

In this case, $g(y) = \exp(-ny - \frac{n}{2} \exp(-2y))$, $t(Y) = \log s$ and $\psi(\theta) = \log \sigma$. Hence, arguing as above we would put a flat prior on $\log \sigma$, i.e. $p(\log \sigma) \propto constant$, which implies that $p(\sigma) \propto \frac{1}{\sigma}$.

When both θ and σ are unknown, we argue that knowledge of θ should be independent of σ and vice versa. Thus, $p(\theta, \sigma^2) = p(\theta)p(\sigma^2) \propto constant \times \frac{1}{\sigma^2} \propto \frac{1}{\sigma^2}$.

More generally, it may be difficult or impossible to write the likelihood in a data-translated format. Jeffreys (1961) suggests the following approach (called *Jeffreys' rule*). Let $\hat{\theta}$ be the mode of the likelihood. The log-likelihood $l(\theta \mid Y)$ is approximately

$$l(\hat{\theta} \mid Y) + \left. \frac{\partial l(\theta \mid Y)}{\partial \theta} \right|_{\hat{\theta}} (\theta - \hat{\theta}) + \frac{(\theta - \hat{\theta})^2}{2} \left[\frac{\partial^2 l(\theta \mid Y)}{\partial \theta^2} \right]_{\hat{\theta}},$$

which is equal to

$$l(\hat{\theta} \mid Y) + \frac{(\theta - \hat{\theta})^2}{2} \left[\frac{\partial^2 l(\theta \mid Y)}{\partial \theta^2} \right]_{\hat{\theta}}$$

$$= l(\hat{\theta} \mid Y) - \frac{n(\theta - \hat{\theta})^2}{2} \left[-\frac{1}{n} \frac{\partial^2 l(\theta \mid Y)}{\partial \theta^2} \right]_{\hat{\theta}}.$$

Note that if

$$\left. \frac{\partial^2 l(\theta \mid Y)}{\partial \theta^2} \right|_{\hat{\theta}}$$

is constant (e.g., equal to $-n\kappa_2$), then the likelihood is approximately data-translated, since

$$L(\theta \mid Y) \approx \exp\left(-\frac{n(\theta - \hat{\theta})^2 \kappa_2}{2} \right),$$

and we would put a flat prior on θ, i.e. $p(\theta) \propto constant$.

Now let $\tau = \tau(\theta)$ be a 1–1 transformation so that $\hat{\tau} = \tau(\hat{\theta})$. Then by the chain rule,

$$\frac{\partial l(\tau \mid Y)}{\partial \tau} = \frac{\partial l(\theta \mid Y)}{\partial \theta} \frac{\partial \theta}{\partial \tau},$$

which implies that

$$-\frac{1}{n} \left(\frac{\partial l(\tau \mid Y)}{\partial \tau} \right)^2 = -\frac{1}{n} \left(\frac{\partial l(\theta \mid Y)}{\partial \theta} \right)^2 \left(\frac{\partial \theta}{\partial \tau} \right)^2.$$

Let

$$J(\tau) = -\frac{1}{n} E\left(\frac{\partial^2 l(\tau \mid Y)}{\partial \tau^2} \right).$$

Since

$$E\left(\frac{\partial^2 l(\tau \mid Y)}{\partial \tau^2} \right) = E\left(\frac{\partial l(\tau \mid Y)}{\partial \tau} \right)^2$$

(see Lee, 1992, page 91), we have

$$J(\tau) = -\frac{1}{n} E\left(\frac{\partial l(\theta \mid Y)}{\partial \theta} \right)^2 \left(\frac{\partial \theta}{\partial \tau} \right)^2 = -\frac{1}{n} E\left(\frac{\partial^2 l(\theta \mid Y)}{\partial \theta^2} \right) \left(\frac{\partial \theta}{\partial \tau} \right)^2$$

$$= J(\theta) \left(\frac{\partial \theta}{\partial \tau} \right)^2.$$

If we choose $\tau(\theta)$ so that $|\frac{\partial\theta}{\partial\tau}| \propto J^{-1/2}(\theta)$, then $J(\hat{\tau}) = \kappa_2$. Now in large samples,

$$I(\hat{\tau}) = -\frac{1}{n}\frac{\partial^2 l(\tau \mid Y)}{\partial\tau^2}\Big|_{\hat{\tau}} \approx E\left[-\frac{1}{n}\frac{\partial^2 l(\tau \mid Y)}{\partial\tau^2}\right]_{\hat{\tau}} = J(\hat{\tau}).$$

In this way, the loglikelihood $l(\tau \mid Y) \approx \kappa_1 - n(\tau - \hat{\tau})^2\kappa_2/2$ and the normalized likelihood is then approximately

$$N\left(\hat{\tau}, \frac{\kappa_2^{-1}}{n}\right).$$

In other words, if we take $|\frac{\partial\theta}{\partial\tau}| \propto J^{-1/2}(\theta)$, then the likelihood *on the τ scale* is approximately data-translated. Arguing as above, we would put a flat prior on $\tau(\theta)$, i.e. $p(\tau(\theta)) \propto constant$ which implies that $p(\theta) \propto |\frac{\partial\tau}{\partial\theta}| \propto J^{1/2}(\theta)$. Note that $J(\theta)$ may not always exist.

Box and Tiao (1973) question why one's ignorance should depend on the given experiment (i.e. that Jeffreys' prior depends on the form of the likelihood). They answer that this noninformative prior expresses ignorance *relative* to the information provided by the intended experiment.

A multivariate extension of Jeffreys' rule would be to take the prior proportional to $(\det J(\theta))^{1/2}$. Jeffreys notes that this rule must be applied with caution—especially where location and scale parameters occur simultaneously (see Exercise 5). Welch and Peers (1963), Berger and Bernardo (1992), and Clarke and Wasserman (1993), among many others, discuss noninformative priors. Kass and Wasserman (1993) present a nice overview of formal rules for selecting priors.

EXAMPLE. *Binomial Distribution*

Under the binomial distribution, the likelihood is $\binom{n}{y}\theta^y(1 - \theta)^{n-y}$ and the loglikelihood is $c + y \log \theta + (n - y) \log(1 - \theta)$. Thus,

$$-\frac{1}{n}\frac{\partial^2 l(\theta \mid Y)}{\partial\theta^2} = -\frac{1}{n}\left[\frac{-y}{\theta^2} - \frac{n - y}{(1 - \theta)^2}\right].$$

Since $Ey = n\theta$, we have

$$J(\theta) = E\left[-\frac{1}{n}\frac{\partial^2 l(\theta \mid Y)}{\partial\theta^2}\right] = \frac{1}{\theta(1 - \theta)}.$$

It then follows that $p(\theta) \propto [\theta(1 - \theta)]^{-1/2} \propto \text{Beta}(1/2, 1/2)$.

As a brief historical note, the Reverend Thomas Bayes (after whom this field is named) considers this case in his paper "Essay Towards Solving a Problem in the Doctrine of Chances," which was published posthumously in the *Philosophical Transactions of the Royal Society (London)* in 1763. It is interesting to note that he takes $p(\theta) \propto constant = \text{Beta}(1,1)$—though concern over this point may have been a factor in Bayes' not submitting the work for publication. (The paper was submitted by Bayes' friend Richard Price after Bayes' death.) In practice, with enough data the choice of Beta(1/2,1/2) versus Beta(1,1) versus Beta(0,0) as a prior will not make much difference. For future reference we will refer to the Beta(0,0) prior as the Haldane prior—see Haldane (1948).

2.2.2. Conjugate Priors

In some situations, the data analyst may not wish to be vague and may be willing to use some smooth distribution that is a specific member of a family of distributions. One such class of distributions is the *conjugate class*.

If the likelihood is based on a random sample of size n from an exponential family distribution with density

$$g(\theta)h(y) \exp \left\{ \sum_{j=1}^{m} \phi_j(\theta)t_j(y) \right\},$$

then a conjugate prior density of the form

$$p(\theta) \propto [g(\theta)]^b \exp \left\{ \sum_{j=1}^{m} \phi_j(\theta)a_j \right\} \qquad (*)$$

combines with the likelihood to yield a posterior density having the same form as $(*)$, but with $b, a_1, \ldots, a_m$ replaced by

$$b' = b + n, \quad a'_j = a_j + \sum_{i=1}^{m} t_j(y_i), \quad j = 1, \ldots, m.$$

For example, the beta distribution is conjugate to the binomial, since adopting the prior distribution $p(\theta) \propto \theta^{a-1}(1 - \theta)^{b-1} \propto \text{Beta}(a, b)$ in conjunction with the likelihood $L(\theta \mid Y) = \theta^y(1 - \theta)^{n-y}$ yields the posterior $p(\theta \mid Y) \propto \theta^{a+y-1}(1 - \theta)^{n-y+b-1} \propto \text{Beta}(a + y, n - y + b)$. Note that the Beta$(a, b)$ prior contributes $a - 1$ successes and $b - 1$ failures, i.e. this prior has the interpretation of additional data.

In a similar way, it can be seen that the gamma distribution is conjugate to the Poisson, the normal distribution is conjugate to the normal (with known σ and unknown μ), and the inverse gamma (IG) is conjugate to the normal (unknown σ and known μ). Note that the noninformative prior for this latter case is $p(\sigma^2) \propto \frac{1}{\sigma^2}$, which corresponds to $IG(0,0)$.

As noted in Smith et al. (1985), there are two limitations to this conjugate prior approach. First, only a limited number of sampling models belong to the exponential family. Second, even when working in an exponential family framework, it may be unrealistic to attempt to represent prior information in the conjugate family form. In this regard, Diaconis and Ylvisaker (1979) expand the notion of conjugate priors and consider mixtures of conjugate priors.

2.3 Maximum Likelihood

Definition 2.3.1. A maximum likelihood estimate (MLE) of θ is a value of θ that maximizes the likelihood $L(\theta \mid Y)$ or, equivalently, the loglikelihood $l(\theta \mid Y)$.

In situations where the maximum likelihood estimate $(\hat{\theta})$ is unique, $\hat{\theta}$ is the value of θ best supported by the data. Alternatively, $\hat{\theta}$ is the value of θ which makes the observed value of the data most likely. It is noted that if $g(\theta)$ is a 1–1 function of θ and $\hat{\theta}$ is an MLE of θ, then $g(\hat{\theta})$ is an MLE of $g(\theta)$.

In this book, we will tend to view $\hat{\theta}$ as a numerical summary of the likelihood— deemphasizing its behavior in repeated samples. The quantity $\hat{\theta}$ is a measure of central tendency. The mode of the likelihood $(\hat{\theta})$ tells us where the likelihood is located.

Suppose the likelihood is differentiable, unimodal, bounded above, and θ is of dimension d. Then the mode $\hat{\theta}$ is obtained by

1. Differentiating the loglikelihood with respect to θ to obtain the $d \times 1$ vector $\partial l(\theta \mid Y)/\partial\theta$.
2. Setting the d simultaneous *likelihood equations* to 0, i.e. $\partial l(\theta \mid Y)/\partial\theta = 0$.
3. Solving for θ.

When the loglikelihood is quadratic, the likelihood equations are linear in θ and the solution is relatively easy to obtain. In more general situations where a closed-form solution of the likelihood equations cannot be found, one may use an iterative numerical method to locate a mode. One such algorithm, the Newton– Raphson algorithm, is described below. Gelman and Rubin (1992) suggest using mode-finding algorithms, such as Newton–Raphson, started at random points, to locate the modes of the posterior density. This information is then used in more complex schemes to summarize the posterior—see Section 6.1.

2.3.1. Newton–Raphson

To motivate the Newton–Raphson algorithm, consider the Taylor series expansion of $l(\theta \mid Y)$ about $\theta^{(i)}$:

$$l(\theta \mid Y) = l(\theta^{(i)} \mid Y) + (\theta - \theta^{(i)})^T \left.\frac{\partial l(\theta \mid Y)}{\partial\theta}\right|_{\theta^{(i)}}$$

$$+ \frac{1}{2}(\theta - \theta^{(i)})^T \left.\frac{\partial^2 l(\theta \mid Y)}{\partial\theta^2}\right|_{\theta^{(i)}} (\theta - \theta^{(i)}) + \text{remainder.}$$

When θ is close to $\theta^{(i)}$, the remainder is negligible. The stationary point of the quadratic approximation

$$l(\theta^{(i)} \mid Y) + (\theta - \theta^{(i)})^T \left.\frac{\partial l(\theta \mid Y)}{\partial\theta}\right|_{\theta^{(i)}} + \frac{1}{2}(\theta - \theta^{(i)})^T \left.\frac{\partial^2 l(\theta \mid Y)}{\partial\theta^2}\right|_{\theta^{(i)}} (\theta - \theta^{(i)})$$

is given by

$$\frac{\partial^2 l(\theta \mid Y)}{\partial \theta^2}\bigg|_{\theta^{(i)}} (\theta - \theta^{(i)}) = -\frac{\partial l(\theta \mid Y)}{\partial \theta}\bigg|_{\theta^{(i)}},$$

which implies

$$\theta = \theta^{(i)} - \left[\frac{\partial^2 l(\theta \mid Y)}{\partial \theta^2}\bigg|_{\theta^{(i)}}\right]^{-1} \frac{\partial l(\theta \mid Y)}{\partial \theta}\bigg|_{\theta^{(i)}}.$$

This suggests the following scheme. Let $\theta^{(0)}$ denote an initial guess of $\hat\theta$ and let $\theta^{(i)}$ denote the guess at the ith iteration. The Newton–Raphson algorithm is given as

$$\theta^{(i+1)} = \theta^{(i)} + \left(-\frac{\partial^2 l(\theta \mid Y)}{\partial \theta^2}\bigg|_{\theta^{(i)}}\right)^{-1} \left[\frac{\partial l(\theta \mid Y)}{\partial \theta}\bigg|_{\theta^{(i)}}\right].$$

If the loglikelihood is a quadratic function of θ, then convergence is obtained after one iteration. If the loglikelihood is concave and unimodal then the sequence $\theta^{(1)}, \theta^{(2)}, \ldots$ converges to $\hat\theta$. The Newton–Raphson algorithm has a quadratic convergence rate near the solution. A poor starting point, however, can lead to divergence—see Exercise 9. An alternative approach, the method of scoring, is given by

$$\theta^{(i+1)} = \theta^{(i)} + \left[E\left(\frac{-\partial^2 l(\theta \mid Y)}{\partial \theta^2}\bigg|\theta\right)\bigg|_{\theta^{(i)}}\right]^{-1} \left\{\frac{\partial l(\theta \mid Y)}{\partial \theta}\bigg|_{\theta^{(i)}}\right\}.$$

For future reference, we will let $S(\theta \mid Y) = \partial l(\theta \mid Y)/\partial \theta$ and refer to $S(\theta \mid Y)$ as the score function. Cox and Hinkley (1974) present an example where $\hat\theta$ is not a solution of the likelihood equations. They present another example, where $\hat\theta$ is not unique. In general, the loglikelihood may have multiple modes and/or saddle points.

2.3.2. Examples

a. Univariate Normal Data

The loglikelihood is given by

$$l(\mu, \sigma^2 \mid Y) = \frac{-n}{2} \ln \sigma^2 - \frac{1}{2} \sum_{i=1}^{n} \frac{(y_i - \mu)^2}{\sigma^2}.$$

The likelihood equations are given by

$$\frac{\partial l}{\partial \mu} = \sum_{i=1}^{n} y_i - n\mu = 0$$

$$\frac{\partial l}{\partial \sigma^2} = \frac{\sum_{i=1}^{n}(y_i - \mu)^2}{2\sigma^4} - \frac{n}{2\sigma^2}.$$

Hence, $\hat{\mu} = \bar{y}$ and $\hat{\sigma}^2 = \Sigma_{i=1}^n (y_i - \bar{y})^2/n$. It remains to show that this stationary point is a maximum using the second derivative of the loglikelihood.

b. Logistic Regression

The data in Table 2.1 are taken from Mendenhall et al. (1989). The first column represents the number of days of radiotherapy received by each of 24 patients. The second column represents the absence (1) or presence (0) of disease at a site three years after treatment. A problem of interest is to use the covariate (days) to predict outcome at three years.

To begin, we wish to fit a logistic regression model (McCullagh and Nelder, 1989) to the data:

$$\log \left(\frac{P_i}{1 - P_i} \right) = \alpha + \beta x_i,$$

where x_i is the covariate for patient i and P_i is the corresponding probability of success (no disease). The logit model implies $P_i = e^{\alpha+\beta x_i}/(1 + e^{\alpha+\beta x_i})$. This model specifies that the log-odds of success is linearly related to the number of days the subject received radiotherapy. The intercept α represents the log-odds of success for zero days, while the slope β represents the change in the log-odds of success for every unit increase in the covariate.

The loglikelihood is given as

$$\sum_{i=1}^{24} \left[y_i \log(P_i) + (1 - y_i) \log(1 - P_i) \right] =$$

TABLE 2.1.

Days	Response	Days	Response
21	1	51	1
24	1	55	1
25	1	25	0
26	1	29	0
28	1	43	0
31	1	44	0
33	1	46	0
34	1	46	0
35	1	51	0
37	1	55	0
43	1	56	0
49	1	58	0

Mendenhall et al. (1989).

$$\sum_{i=1}^{24}\left[y_i(\alpha+\beta x_i)-\log(1+e^{\alpha+\beta x_i})\right].$$

Note that

$$\frac{\partial l(\theta\mid Y)}{\partial\alpha}=\sum_{i=1}^{24}\left(y_i-\frac{e^{\alpha+\beta x_i}}{1+e^{\alpha+\beta x_i}}\right)=\sum_{i=1}^{24}(y_i-P_i)$$

$$\frac{\partial l(\theta\mid Y)}{\partial\beta}=\sum_{i=1}^{24}\left(y_ix_i-\frac{x_ie^{\alpha+\beta x_i}}{1+e^{\alpha+\beta x_i}}\right)=\sum_{i=1}^{24}x_i(y_i-P_i)$$

and

$$\frac{\partial^2 l(\theta\mid Y)}{\partial\alpha^2}=-\sum_{i=1}^{24}\frac{e^{\alpha+\beta x_i}}{(1+e^{\alpha+\beta x_i})^2}=-\sum_{i=1}^{24}P_i(1-P_i)$$

$$\frac{\partial^2 l(\theta\mid Y)}{\partial\alpha\partial\beta}=-\sum_{i=1}^{24}\frac{x_ie^{\alpha+\beta x_i}}{(1+e^{\alpha+\beta x_i})^2}=-\sum_{i=1}^{24}x_iP_i(1-P_i)$$

$$\frac{\partial^2 l(\theta\mid Y)}{\partial\beta^2}=-\sum_{i=1}^{24}\frac{x_i^2 e^{\alpha+\beta x_i}}{(1+e^{\alpha+\beta x_i})^2}=-\sum_{i=1}^{24}x_i^2 P_i(1-P_i).$$

Since $\partial^2 l(\theta\mid Y)/\partial\theta^2$ does not depend on Y, the Newton–Raphson algorithm and the method of scoring are equivalent for this model.

Implementing the Newton–Raphson algorithm for this data set, we find that $\hat{\alpha}=3.819$ and $\hat{\beta}=-0.087$ after four iterations from the starting point (0.4, 0.4). Over the range of the data, the effect of an additional day of treatment is to *decrease* the odds of success at 3 years by about 8%.

c. Item Response Model

Suppose n subjects each take an examination with k items. Each item is scored as correct or incorrect and Y_{ij} equals 1 if the ith student correctly answers the jth item and 0 otherwise. Let Y_{ij} be independent Bernoulli random variables where the probability of success is

$$p_{ij}=P(Y_{ij}=1\mid\theta_i,\alpha_j,\beta_j)=\frac{\exp(\alpha_j\theta_i-\beta_j)}{1+\exp(\alpha_j\theta_i-\beta_j)},$$

i.e. a two-parameter logistic model. The quantity θ_i represents the unobserved "ability" of the ith student, while α_j measures the discriminatory power (to discriminate between strong and weak students) of the jth item and β_j represents the difficulty of the jth item. It is standard (see Albert, 1992, Tsutakawa, 1984 or Tsutakawa and Lin, 1986) to require that $\alpha_j>0$ so that under the model stronger students are *more* likely to correctly answer the jth item.

The likelihood is given by

$$\prod_{i=1}^{n}\prod_{j=1}^{k}p_{ij}^{y_{ij}}(1-p_{ij})^{1-y_{ij}},$$

assuming conditional independence among the responses. Regarding the prior, Albert (1992), for example, assumes that $\theta_1, \ldots, \theta_n$ are an *iid* sample from a $N(0, 1)$ population and he puts an *iid* flat prior on $\zeta_j = (\alpha_j, \beta_j)$, $j = 1, \ldots, k$. Under this specification, the posterior is proportional to

$$\prod_{i=1}^{n} \phi(\theta_i) \prod_{j=1}^{k} p_{ij}^{y_{ij}} (1 - p_{ij})^{1 - y_{ij}},$$

where $\phi(.)$ is the standard normal density. Since n can be quite large and the abilities $\theta_1, \ldots, \theta_n$ are considered nuisance parameters, it has been suggested that one base inference on the marginal

$$p(\zeta \mid data) = \int p(\theta, \zeta \mid data) d\theta,$$

where $\zeta = (\zeta_1, \ldots, \zeta_k)$ and $\theta = (\theta_1, \ldots, \theta_n)$, thereby reducing the dimension of the parameter space from $n + 2k$ to $2k$. Note that the logarithm of the marginal posterior is

$$\log(p(\zeta \mid data)) = \sum_{i=1}^{n} \log \left[\int \phi(\theta_i) \prod_{j=1}^{k} p_{ij}^{y_{ij}} (1 - p_{ij})^{1 - y_{ij}} d\theta_i \right].$$

Differentiating the log-marginal with respect to α_j and β_j, for $j = 1, \ldots, k$, we obtain the equations:

$$\sum_{i=1}^{n} y_{ij} = \sum_{i=1}^{n} \int \frac{\exp(\alpha_j \theta_i - \beta_j) p(\theta_i \mid Y, \zeta)}{1 + \exp(\alpha_j \theta_i - \beta_j)} d\theta_i$$

$$\sum_{i=1}^{n} y_{ij} \int \theta_i p(\theta_i \mid Y, \zeta) d\theta_i = \sum_{i=1}^{n} \int \frac{\theta_i \exp(\alpha_j \theta_i - \beta_j) p(\theta_i \mid Y, \zeta)}{1 + \exp(\alpha_j \theta_i - \beta_j)} d\theta_i,$$

for $j = 1, \ldots, k$, where $p(\theta_i \mid Y, \zeta)$ is proportional to $\phi(\theta_i) \prod_{j=1}^{k} p_{ij}^{y_{ij}} (1 - p_{ij})^{1 - y_{ij}}$.

The item response model can be viewed as a logistic regression model with a random covariate. Note that the terms P_i and x_i in the logistic regression likelihood equations are replaced by their expectations in the above equations for the item response model. Unfortunately, since the vector ζ appears in each of the k sets of equations, we end up with $2k$ unknowns in $2k$ equations. As we will see in the *EM* solution for the item response model, the calculations reduce to solving k independent sets of 2 equations in 2 unknowns. The Gibbs sampler solution further simplifies matters by providing a sampling based approach for evaluating the (nonanalytically tractable) integrals.

2.4 Normal-Based Inference

Let $\hat{\theta}$ be the mode of the likelihood for the data set Y. Then in certain situations

$$(\theta - \hat{\theta}) \sim N(0, C), \tag{2.4.1}$$

where C is the $d \times d$ variance–covariance matrix of $(\theta - \hat{\theta})$.

To Frequentists, θ is fixed (though unknown), while $\hat{\theta}$ is random. Given the model and the value θ, (2.4.1) shows that in repeated samples $\hat{\theta}$ will be normally distributed with mean θ and variance–covariance matrix C.

To Bayesians and to likelihoodists, $\hat{\theta}$ is fixed (conditional on the data Y) and θ is random. Given the model and the data, (2.4.1) indicates that the posterior density of θ is normal with mean $\hat{\theta}$ and variance–covariance matrix C, i.e. $p(\theta \mid Y) = N(\hat{\theta}, C)$.

The Frequentist justification of (2.4.1) is an expansion of $S(\hat{\theta} \mid Y)$ in a Taylor series about the fixed, but unknown constant, θ. Under some regularity conditions, asymptotically,

$$J(\theta)(\theta - \hat{\theta}) \approx S(\theta \mid Y),$$

where $J(\theta) = E(-\partial^2 L(\theta \mid Y)/\partial\theta^2 \mid \theta)$ and $J(\theta)$ is called the *expected information matrix*. Thus, $(\hat{\theta} - \theta) \approx J^{-1}(\theta)S(\theta \mid Y)$. Under certain regularity conditions, in repeated samples $S(\theta \mid Y)$ is asymptotically normal with mean 0 and variance–covariance matrix $J(\theta)$. It follows that $(\theta - \hat{\theta})$ is normally distributed with mean 0 and variance–covariance matrix $C = J^{-1}(\theta)$. We evaluate $J^{-1}(\theta)$ at $\hat{\theta}$ and ignore the resulting error.

It is important to note that $J^{-1}(\theta)$ is the variance of the asymptotic distribution, not the limit of the exact variance.

The Bayesian justification of (2.4.1) is an expansion of the loglikelihood (log-posterior) about the fixed value $\hat{\theta}$ (conditional on Y),

$$l(\theta \mid Y) = l(\hat{\theta} \mid Y) + (\theta - \hat{\theta})^T S(\hat{\theta} \mid Y) - \frac{1}{2}(\theta - \hat{\theta})^T I(\hat{\theta} \mid Y)(\theta - \hat{\theta}) + r(\theta \mid Y),$$

where $I(\hat{\theta} \mid Y) = -\partial^2 l(\theta \mid Y)/\partial\theta^2|_{\hat{\theta}}$ and $I(\theta \mid Y)$ is called the *observed information matrix*. Since $S(\hat{\theta} \mid Y) = 0$, assuming that $r(\theta \mid Y)$ can be neglected, the likelihood/posterior density is proportional to the multivariate normal density with mean $\hat{\theta}$ and variance–covariance matrix $C = I^{-1}(\hat{\theta} \mid Y)$.

Based on the normal approximation, a $100\alpha\%$ interval for θ (scalar) is given by $\hat{\theta} \pm zC^{1/2}$, where z is the value such that the area under the normal distribution from $-z$ to $+z$ is $100\alpha\%$, and C is either $I^{-1}(\hat{\theta} \mid Y)$ or $J^{-1}(\hat{\theta})$. Frequentists would have $100\alpha\%$ confidence that this interval contains the (unknown) value of θ. Bayesians would say that θ has a $100\alpha\%$ chance of being in this interval.

For a vector θ of dimension d, a $100\alpha\%$ ellipsoid is given by

$$(\theta - \hat{\theta})^T C^{-1} (\theta - \hat{\theta}) \le \chi^2,$$

where χ^2 is the value such that the area under the chi-square distribution of d degrees of freedom from 0 to χ^2 is $100\alpha\%$.

Efron and Hinkley (1978) discuss the relative merits of I^{-1} and J^{-1}. In some sense, I^{-1} is preferred since it gives a better indication of the curvature of the *observed* likelihood. In some situations, I^{-1} may be easier to compute than J^{-1}—in other situations the reverse may be true!

EXAMPLE. *Exponential Lifetimes with Random Censoring*
In this case, suppose $T_1, \ldots, T_n$ are *iid* with exponential density $f(t) = \theta e^{-\theta t}$ and survivor function $S(t) = e^{-\theta t}$, $(t > 0)$. Let n_μ be the number of uncensored observations in the sample. The loglikelihood is given as

$$l(\theta \mid Y) = n_\mu \log \theta - \theta \sum_{i=1}^{n} y_i,$$

where y_i is the observed event time. The score function is

$$\frac{n_\mu}{\theta} - \sum_{i=1}^{n} y_i,$$

which implies that $\hat{\theta} = n_\mu / \Sigma_{i=1}^{n} y_i$. Note that $-\partial^2 l / \partial \theta^2 = n_\mu / \theta^2 = I(\theta \mid Y)$, while $J(\theta) = np(T < C)/\theta^2$. Hence, we can approximate the normalized likelihood with a normal with mean $\hat{\theta}$ and variance $\hat{\theta}^2 / n_\mu$. A 95% interval for θ is given

$$\hat{\theta} \pm 1.96 \sqrt{\frac{\hat{\theta}^2}{n_\mu}}.$$

A plot of the normalized likelihood and the normal approximation could be used to assess the adequacy of the normal distribution in capturing the shape of the likelihood.

EXAMPLE. *Logistic Regression (Continued)*
For the logistic regression data in Section 2.3, the upper-left entry in $I^{-1}(\hat{\alpha}, \hat{\beta} \mid Y)$ is 3.345, while the lower-right entry is 0.00185. Hence, the standard error of $\hat{\alpha} = 1.829$ and standard error of $\hat{\beta} = 0.043$. A 95% interval for β obtained via $\hat{\beta} \pm 1.96 \times SE(\hat{\beta})$ is $(-0.171, -0.003)$. Thus, because the interval does not overlap zero, there appears to be evidence of a day effect on the odds of success at 3 years.

EXAMPLE. 2×2 *Table*
Consider the 2×2 table

	Population 1	Population 2
Success	x	y
Failure	m − x	n − y
Total	m	n

where m and n are fixed by design. Suppose that x has a binomial distribution with parameters (m, π) and that y has an independent binomial distribution with parameters (n, ρ). Adopting a Beta(α_o, β_o) prior for π and an independent Beta(γ_o, δ_o) prior for ρ, where α_o, β_o, γ_o and δ_o are known constants, it follows that the posterior distribution for π is Beta$(\alpha_o + x, \beta_o + m - x)$ and the posterior distribution for ρ is Beta$(\gamma_o + y, \delta_o + n - y)$.

Via simulation, it is straightforward to obtain the posterior distribution of the difference $\pi - \rho$ or, for that matter, the posterior distribution of any functional of π and ρ. To obtain an observation from $p(\pi - \rho \mid data)$,

1. Draw π^* from Beta$(\alpha_o + x, \beta_o + m - x)$.
2. Draw ρ^* from Beta$(\gamma_o + y, \delta_o + n - y)$.
3. Compute $\pi^* - \rho^*$.

Repeating steps 1–3 yields an *iid* sample from $p(\pi - \rho \mid data)$.

For large m and n, the posterior distribution of the *log-odds ratio*

$$\log\left(\frac{\pi(1-\rho)}{(1-\pi)\rho}\right) = \log\left(\frac{\pi}{1-\pi}\right) - \log\left(\frac{\rho}{1-\rho}\right)$$

can be approximated by a normal distribution with mean

$$\log\left[\frac{\left(\alpha_o + x - \frac{1}{2}\right)\left(\delta_o + n - y - \frac{1}{2}\right)}{\left(\beta_o + m - x - \frac{1}{2}\right)\left(\gamma_o + y - \frac{1}{2}\right)}\right] \qquad (*)$$

and variance

$$\frac{1}{\alpha_o + x} + \frac{1}{\beta_o + m - x} + \frac{1}{\gamma_o + y} + \frac{1}{\delta_o + n - y}$$

(see Exercise 14). Note that if we adopt a Haldane prior on π and on ρ, i.e. take $\alpha_o = \beta_o = \gamma_o = \delta_o = 0$ and we ignore the 1/2 terms in $(*)$, then we can approximate the posterior distribution of the log-odds ratio

$$\log\left(\frac{\pi(1-\rho)}{(1-\pi)\rho}\right)$$

as a normal distribution with mean

$$\log\left(\frac{x(n-y)}{(m-x)y}\right),$$

i.e. the logarithm of the cross-product ratio (see Haberman, 1978, p. 105), and variance

$$\frac{1}{x} + \frac{1}{m-x} + \frac{1}{y} + \frac{1}{n-y}.$$

2.5 The δ-Method (Propagation of Errors)

Suppose that the distribution of $\theta - \hat{\theta}$ can be approximated by a normal distribution with mean 0 and variance–covariance matrix B. Let f be a function defined on an open subset of a d-dimensional space taking values in an r-dimensional space and let f be differentiable at θ. The approximating distribution to $f(\hat{\theta}) - f(\theta)$ is the normal distribution with mean 0 and variance–covariance matrix $(\partial f/\partial \theta)^T B (\partial f/\partial \theta)|_{\hat{\theta}}$.

EXAMPLE. *Asymptotic Variance of the Log Odds Ratio*
Consider the log odds ratio for a multinomially distributed 2×2 table with observed proportions P_{ij}, which is given by

$$\log\left(\frac{P_{11}P_{22}}{P_{21}P_{12}}\right) = \log(P_{11}) + \log(P_{22}) - \log(P_{21}) - \log(P_{12}).$$

Note that $B = (D - P^T P)/N$, where D is the diagonal matrix with vector $P = (P_{11}, P_{22}, P_{12}, P_{21})$ along its main diagonal and N is the total sample size. Also note that $\partial f/\partial P = CD^{-1}$, where $C = [1, 1, -1, -1]$. The variance of the log odds ratio is then given by

$$[CD^{-1}(D - P^T P)D^{-1}C^T]/N$$
$$= [CD^{-1}C^T - (CD^{-1}P^T)PD^{-1}C^T]/N$$
$$= \left(\frac{1}{P_{11}} + \frac{1}{P_{22}} + \frac{1}{P_{12}} + \frac{1}{P_{21}}\right)\Big/N.$$

See Bishop, Fienberg and Holland (1975) for further examples of the use of the δ-method.

2.6 Highest Posterior Density Regions

From the Bayesian perspective, inferential problems about the parameter vector θ can be addressed via the posterior density $p(\theta \mid Y)$. However, when θ is of even moderate dimension, e.g. six, it may be difficult to grasp its shape. Thus, methods which summarize the structure of the posterior are of value. The Highest Posterior

Density region attempts to capture a comparatively small region of the parameter space which contains most of the mass of the posterior distribution. Intuitively, the HPD region is motivated as the region where the probability density of every point inside the region is at least as large as that of any point outside the region. More formally, we have:

Definition 2.6. (Box and Tiao, 1973) A region R in parameter space Θ is called the highest posterior density region of content α if

a. $\Pr(\theta \in R \mid Y) = \alpha$
b. For $\theta_1 \in R, \theta_2 \notin R, p(\theta_1 \mid Y) \geq p(\theta_2 \mid Y)$.

Thus, consider the two regions R and R' (see Figure 2.1). R is an HPD region of content α, while R' is not.

Box and Tiao (1973) point out that for a given probability content α, the HPD region has smallest volume in parameter space. In addition, θ_o is covered by the HPD region of content α iff

$$P_\theta[p(\theta \mid Y) \geq p(\theta_o \mid Y) \mid Y] \leq \alpha$$

where $p(\theta \mid Y)$ is treated as a random variable.

EXAMPLE. *(Linear Model):*
Consider the linear model given as

$$y = x\theta + \varepsilon,$$

where y is an $n \times 1$ vector of observations, x is an $n \times d$ matrix of known constants, θ is a $d \times 1$ vector of unknown regression coefficients, ε is an $n \times 1$ vector of independent errors, where $\varepsilon_i \sim N(0, \sigma^2)$ and σ is unknown.

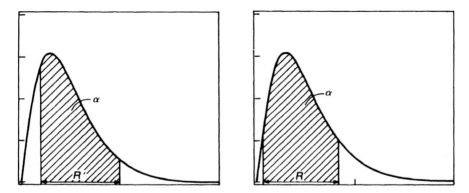

FIGURE 2.1.

If we define $Q(\theta) = (\theta - \hat{\theta})^T (x^T x)(\theta - \hat{\theta})$, then

a. The posterior of $Q(\theta)/ds^2$ is $F(d, v)$, where $s^2 = (y - \hat{y})^T (y - \hat{y})/v$, $v = n - d$, and $\hat{y} = x\hat{\theta}$.

b. The posterior density $p(\theta \mid Y)$ is a monotonically decreasing function of $Q(\theta)$.

In this way, the content of the smallest HPD region containing θ_o is equal to

$$P_\theta[p(\theta \mid Y) \geq p(\theta_o \mid Y) \mid Y]$$

$$= P_\theta[(\theta_o - \hat{\theta})^T x^T x(\theta_o - \hat{\theta}) \geq (\theta - \hat{\theta})^T x^T x(\theta - \hat{\theta}) \mid Y]$$

$$= \Pr\left[\frac{(\theta_o - \hat{\theta})^T x^T x(\theta_o - \hat{\theta})}{ds^2} \geq F(d, v) \right]$$

$$= 1 - \text{Frequentist significance level for testing } \theta = \theta_o.$$

In this small sample context, the Bayesian (under a noninformative prior) and the Frequentist approaches yield the same results. Box and Tiao (1973) point out that these results supply a Bayesian justification for the analysis of variance. More generally, Wasserman (1989) provides a Bayesian interpretation to the likelihood region. He proves that likelihood regions are robust in the sense that their posterior probability content is relatively insensitive to contaminations of the prior. Datta and Ghosh (1995) derive a differential equation that a prior must satisfy if the posterior probability of a one-sided credible interval for a parameter and its Frequentist probability agree to $0\left(\frac{1}{n}\right)$.

Exercises

1. Consider an *iid* sample of size n from the $N(\mu, \sigma^2)$ distribution, where σ^2 is known. Derive the posterior distribution of μ under the prior $N(\mu_o, \sigma_o^2)$, where μ_o and σ_o are known constants.

2. Charles Darwin in his book *The Effects of Cross and Self Fertilisation in the Vegetable Kingdom* examines the advantages of cross-fertilization in plants. He reports the following data on the difference in heights in corn plants between paired crossed and self-fertilized plants. The data are in eighths of an inch.

   49, −67, 8, 16, 6, 23, 28, 41, 14, 29, 56, 24, 75, 60, −48

   Suppose that the data are normally distributed with mean μ and variance σ^2 (unknown). Under the noninformative prior, derive the posterior of μ. What is your best guess for μ? Obtain a 95% HPD interval for μ and interpret this interval. Compute $p(\mu > 0 \mid Y)$.

3. The following data, taken from Wallace (1980), represent the hours of post-operative pain relief for subjects receiving one of two drugs:

drug #1: 2, 4, 4, 5, 6, 8, 13
drug #2: 0, 0, 0, 1, 1, 2, 2, 2, 3, 3, 3, 4, 8

  a. Assuming that the data are normally distributed and $\sigma_1^2 = \sigma_2^2$ (unknown), under independent noninformative priors compute the posterior distribution of $\mu_1 - \mu_2$. Obtain a 95% HPD interval. Compute $p(\mu_1 - \mu_2 > 0 \mid Y)$.

  b. Do not assume that $\sigma_1^2 = \sigma_2^2$. Via simulation, repeat (a). (*Hint:* Draw independently from the t distribution on 6 df and from the t distribution on 12 df.) How would you quantify the Monte Carlo error in your computations?

  c. Again, do not assume that $\sigma_1^2 = \sigma_2^2$. Using the Patil approximation to the Behrens–Fisher distribution, repeat (a).

  d. Using the Welch t approximation, compute a 95% confidence interval for $\mu_1 - \mu_2$. Compute the area under the posterior over the interval specified by the Welch approach. Is the area close to 95%?

  e. Compare your results from parts (a), (b), (c) and (d). Of particular interest is the location and length of the intervals.

4. Suppose that $x_1, \ldots, x_n$ is an *iid* sample from the $N(\mu, \sigma_x^2)$ and $y_1, \ldots, y_m$ is an *iid* sample from the $N(\lambda, \sigma_y^2)$, independent of the first sample, with μ, σ_x^2, λ and σ_y^2 all unknown. Derive the posterior distribution of $\frac{\sigma_x^2/s_x^2}{\sigma_y^2/s_y^2}$ under independent noninformative priors, where s_x^2 is the sample variance of the x's and s_y^2 is the sample variance of the y's.

5. Let $y_1, \ldots, y_n$ be an *iid* sample from the Poisson distribution with parameter λ. Derive Jeffrey's (noninformative) prior. This prior corresponds to a gamma distribution with which parameters?

6. In the multivariate setting, $\theta = (\theta_1, \ldots, \theta_d)$, $p(\theta) \propto \mid J(\theta) \mid^{1/2}$ provides an invariant prior, where the ijth entry of $J(\theta)$ is equal to

$$-E \left(\frac{\partial^2 l(\theta \mid Y)}{\partial \theta_i \partial \theta_j} \right)$$

and $\mid X \mid$ is the determinant of the matrix X.

    Let $y_1, \ldots, y_n$ be an *iid* sample from the $N(\mu, \sigma^2)$ distribution, where μ and σ are both unknown. Derive the invariant prior. How does it compare with the prior $p(\theta, \sigma^2) \propto 1/\sigma^2$?

7. Let p denote the probability that a specific major league baseball player will get a hit in a particular time at bat. Assume that batting averages usually fall in the range .19 to .35.

  a. Consider the priors Uniform (.19, .35); Beta(10.2, 23.8); and Beta(20.4, 47.6). Plot these priors and discuss each choice.

  b. Suppose a player gets 5 hits in 40 at-bats. For each of the above priors: plot the likelihood, posterior and prior; compute the posterior mean and posterior standard deviation; compute the probability that the player is better than a .200 hitter; compute your best guess as to the batting average of the player; and compute a 95% HPD interval for p.

c. In general, for a binomial likelihood and beta prior, show that the posterior mean lies between the prior mean and the proportion of successes.

8. The following data represent the number of arrivals for 45 time intervals of length 2 minutes at a cashier's desk at a supermarket and are taken from Andersen (1980):

0, 0, 0, 0, 0, 0, 1, 1, 1, 1, 1, 1, 1, 1, 1, 1, 1, 1, 1, 1, 1, 1, 1, 1, 1,
2, 2, 2, 2, 2, 2, 2, 2, 2, 3, 3, 3, 3, 3, 3, 3, 4, 4, 4, 4, 5

a. For a Gamma $(2, 1)$ prior, obtain the posterior distribution under a Poisson (λ) model for the data. Draw the prior and the posterior. Note on your plot the mean, standard deviation and mode of the posterior. Is the prior sensible? What "information" does it add to the data?

b. For the noninformative prior, i.e. a Gamma $(1/2, 0)$, repeat part (a).

9. 197 animals are distributed into four categories: $Y = (y_1, y_2, y_3, y_4)$ according to the genetic linkage model $((2 + \theta)/4, (1-\theta)/4, (1-\theta)/4, \theta/4)$.

a. What is the likelihood for the data $Y = (125, 18, 20, 34)$?

b. What is the likelihood for the data $Y = (14, 0, 1, 5)$?

c. Use the Newton–Raphson algorithm to obtain the MLE $(\hat{\theta})$ of θ for $Y = (125, 18, 20, 34)$. Try starting your algorithm at $\theta = .1, .2, .3, .4, .6$, and .8. How did you assess convergence of the algorithm?

d. Repeat (c) for $Y = (14, 0, 1, 5)$. Did the algorithm converge for all the starting points?

e. Plot the normalized likelihood and the associated normal approximation in the same figure for the data $Y = (125, 18, 20, 34)$. Discuss the adequacy of the normal approximation.

f. Repeat (e) for $Y = (14, 0, 1, 5)$.

10. Suppose a sample of size n is drawn at random and with replacement from an urn containing white balls and black balls. For large n, the posterior distribution of $\hat{p} - p$, where p is the true proportion of white balls in the urn and $\hat{p}$ is the proportion of white balls in the sample, is normal with mean 0 and variance $\frac{p(1-p)}{n}$. Find the asymptotic distribution of $2 \sin^{-1} \sqrt{\hat{p}} - 2 \sin^{-1} \sqrt{p}$ using the delta method.

11. Let $x_1, \ldots, x_n$ be an iid sample from $N(\theta, 1)$ and let $y_1, \ldots, y_n$ be an independent iid sample from $N(\phi, 1)$. Derive the asymptotic distribution of $\theta/\phi - \bar{x}/\bar{y}$ ($\phi, \bar{y} \neq 0$) via the delta method.

12. Derive Result 2.1.2.

13. Derive Result 2.1.3.

14. Suppose that π has a Beta(α, β) distribution.

a. What is the distribution of $\lambda = \frac{\beta\pi}{\alpha(1-\pi)}$?

b. Show that $\delta = \frac{1}{2}\log(\lambda)$ has the density

$$p(\delta) = \frac{2(2\alpha)^\alpha (2\beta)^\beta}{Be(\alpha, \beta)} \frac{e^{2\alpha\delta}}{(2\beta + 2\alpha e^{2\delta})^{(\alpha+\beta)}} .$$

This distribution is known as *Fisher's z distribution*.

c. Show that the mean of $p(\delta)$ is approximately

$$\frac{1}{2} \log\left[\left(1 - \frac{1}{2\alpha}\right) \Big/ \left(1 - \frac{1}{2\beta}\right)\right]$$

and that the variance is approximately

$$\frac{1}{4}\left(\frac{1}{\alpha} + \frac{1}{\beta}\right).$$

It then follows that the mean and variance of

$$\log\left(\frac{\pi}{1 - \pi}\right)$$

is

$$\log\left[\left(\alpha - \frac{1}{2}\right) \Big/ \left(\beta - \frac{1}{2}\right)\right]$$

and $\frac{1}{\alpha} + \frac{1}{\beta}$, respectively.

d. If the posterior distribution for π is Beta($\alpha_o + x$, $\beta_o + m - x$) and the (independent) posterior distribution for ρ is Beta($\gamma_o + y$, $\delta_o + n - y$), then what is the approximate mean and variance of the *log-odds ratio*

$$\log\left(\frac{\pi(1 - \rho)}{(1 - \pi)\rho}\right) = \log\left(\frac{\pi}{1 - \pi}\right) - \log\left(\frac{\rho}{1 - \rho}\right)?$$

3

Nonnormal Approximations to Likelihoods and Posteriors

In this chapter, we continue our presentation of methods which are applied directly to the likelihood or to the posterior. However, in this chapter we work with higher-order approximations to these functions. In Section 3.1, we discuss numerical integration as a method to obtain a close approximation to the marginal of the posterior or of the likelihood. Section 3.2 approaches the problem of higher-order approximations to the posterior or to the likelihood from the point of view of Laplace's method. Section 3.3 presents the methods of Monte Carlo, importance sampling and rejection/acceptance to realize a sample from the function of interest. Iterative Monte Carlo methods are presented in chapters 5 and 6.

3.1 Numerical Integration

In the previous chapter, we approximated the posterior or the likelihood by a normal distribution and then used the normal distribution as our basis of inference. Unfortunately, for a given model and a given data set, the normal approximation may be quite poor and may lead to inappropriate inferences. Alternatively, we can integrate out nuisance parameters and use the resulting marginal posterior or likelihood as the basis of inference.

To obtain the marginal density of $(\theta_1, \ldots, \theta_l)$, where $l < d$, we integrate over $(\theta_{l+1}, \ldots, \theta_d)$ to obtain

$$p(\theta_1, \ldots, \theta_l \mid Y) = \int p(\theta_1, \ldots, \theta_d \mid Y) d\theta_{l+1} \ldots d\theta_d.$$

More generally, we may be interested in evaluating the integral

$$\int q(\theta) p(\theta \mid Y) d\theta_I \tag{3.1.1}$$

where I is the complement of the subscripts of the components of θ of interest in $(1, 2, \ldots, d)$.

Smith et al. (1985) and Naylor and Smith (1982) discuss their experience with methods from the numerical analysis literature to integrate out numerically the nuisance parameters. Their approach is to use Cartesian product Gaussian quadrature rules (Davis and Rabinowitz, 1984) for multivariate parameter vectors. Such an approach assumes that the posterior density may be well approximated by the product of a multivariate normal distribution and a polynomial in $\theta_1, \ldots, \theta_k$ and that the parameters are nearly orthogonal. To justify the latter condition, Naylor and Smith (1982) transform the component parameters to a new, orthogonal set of parameters.

More formally, in the one-dimensional case, if $h(t)$ is a suitably regular function and

$$g(t) = h(t)(2\pi\sigma^2)^{-1/2} \exp\left\{-\frac{1}{2}\left[\frac{t-u}{\sigma}\right]^2\right\},$$

then Naylor and Smith (1982) show that

$$\int_{-\infty}^{\infty} g(t)dt \approx \sum_{i=1}^{n} m_i g(z_i), \qquad (3.1.2)$$

where

$$m_i = w_i \exp(t_i^2)\sqrt{2}\sigma, \ z_i = \mu + \sqrt{2}\sigma t_i. \qquad (3.1.3)$$

To implement this approach, tables of t_i, w_i and $w_i \exp(t_i^2)$ are required (Naylor and Smith, 1982). The error in the approximation will be small if $h(t)$ is approximately a polynomial. Maximum likelihood values make reasonable guesses to μ and σ in (3.1.3). In the case of a k-dimensional parameter vector, a simple extension of (3.1.2) is

$$\int \cdots \int g(t_1, \ldots, t_k) \, dt_1 \cdots dt_k \approx$$
$$\sum_{i_k} m_{i_k}^{(k)} \cdots \sum_{i_2} m_{i_2}^{(2)} \sum_{i_1} m_{i_1}^{(1)} g\left(z_{i_1}^{(1)} \cdots z_{i_k}^{(k)}\right),$$

where $m_{i_j}^{(j)}$, $z_{i_j}^{(j)}$ are found via (3.1.3), but estimates of the marginal posterior mean and variance of θ_i substituted for μ and σ^2.

EXAMPLE. *Logistic Regression (Continued)*
Figure 3.1 presents the β marginal (solid line) of the likelihood obtained via numerical integration for the logistic regression data of Section 2.3. In particular, if $L(\alpha, \beta \mid Y)$ is the likelihood, then to obtain $L(\beta^* \mid Y)$ (i.e. the β marginal evaluated at the point β^*) we used the trapezoidal rule:

$$\frac{\alpha_m - \alpha_0}{m}\left[\frac{1}{2} L(\alpha_0, \beta^* \mid Y) + L(\alpha_1, \beta^* \mid Y) + \cdots \right.$$
$$\left. + L(\alpha_{m-1}, \beta^* \mid Y) + \frac{1}{2} L(\alpha_m, \beta^* \mid Y)\right],$$

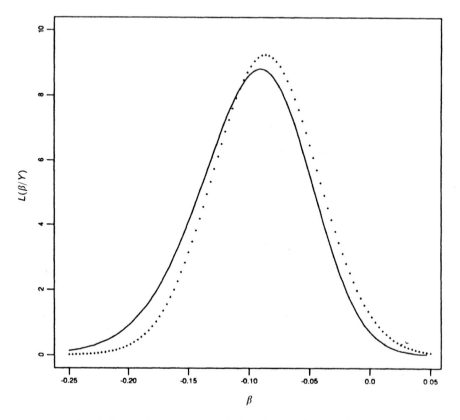

FIGURE 3.1. The beta marginal (solid: exact; dotted: normal approximation). The curves are normalized to integrate to unity.

where $\alpha_0, \ldots, \alpha_m$ defines a grid of equispaced points along the α axis. This computation was repeated for β values over $[-0.25, 0.005]$ to obtain $L(\beta^{(1)} \mid Y)$, $L(\beta^{(2)} \mid Y), \ldots, L(\beta^{(n)} \mid Y)$. The trapezoidal rule was then used to obtain the normalizing constant to yield Figure 3.1. The dotted line represents the normal approximation to the marginal. As can be seen, the true marginal of β is slightly skewed, which is not detected in the normal approximation. The difference is minor in terms of the mode (new mode $= -0.088$). However, the exact HPD region of content 95% is $(-0.185, -0.015)$, rather than $(-0.171, -0.002)$ based on the normal approximation to the likelihood.

Figure 3.2 presents the $-\alpha/\beta$ (i.e. TD_{50}) marginal. The solid line represents the normal approximation obtained via the δ-method. The dotted curve was obtained by reparameterizing the likelihood in terms of $\rho(= -\alpha/\beta)$ and α and then numerically

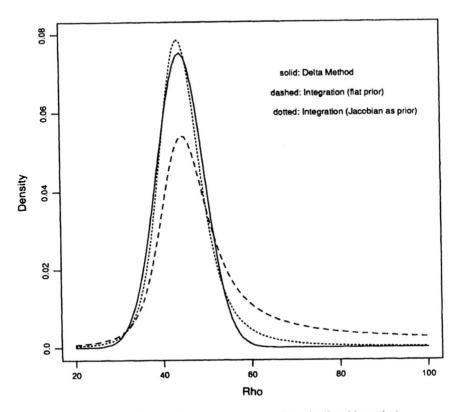

FIGURE 3.2. Delta method vs. integration (flat prior/Jacobian prior).

integrating out α. In this case, the prior on ρ was the Jacobian of the transformation. In other words, the dotted curve was obtained via:

$$\int L(\rho, \alpha \mid Y) \mid \alpha \mid / \rho^2 \, d\alpha.$$

The dashed line represents the marginal having adopted a flat prior on ρ, i.e. via:

$$\int L(\rho, \alpha \mid Y) d\alpha.$$

As can be seen, the mode of each curve is located at about 44 days. The integrated curve (Jacobian prior) is slightly more skewed than the normal approximation. There is evidence that the integrated curve (with a flat prior) is not proper, i.e. has infinite mass. In particular, evaluating the marginal at $-\alpha/\beta = 200, 300, \ldots, 2000$ indicates that the marginal is flat. In this regard, it is noted that the dashed curve in Fig. 3.2 was truncated at $-\alpha/\beta = 100$ and scaled to integrate to 1.0. Truncating at a larger value would scale the curve downward. For related work, see Buonaccorsi and Gatsonis (1988).

3.2 Posterior Moments and Marginalization Based on Laplace's Method

3.2.1. Moments

Suppose $-h(\theta)$ is a smooth, bounded unimodal function, with a maximum at $\hat{\theta}$, and θ is a scalar. By Laplace's method (Tierney and Kadane, 1986), the integral

$$I = \int f(\theta) \exp[-nh(\theta)]d\theta$$

can be approximated by

$$\hat{I} = f(\hat{\theta})\sqrt{\frac{2\pi}{n}} \, \sigma \, \exp[-nh(\hat{\theta})],$$

where

$$\sigma = \left[\frac{\partial^2 h}{\partial \theta^2}\Big|_{\hat{\theta}} \right]^{-1/2}.$$

As presented in Mosteller and Wallace (1964), Laplace's method is to expand about $\hat{\theta}$ to obtain:

$$I \approx \int f(\hat{\theta}) \exp\left\{-n\left[h(\hat{\theta}) + (\theta - \hat{\theta})h'(\hat{\theta}) + \frac{(\theta - \hat{\theta})^2}{2} h''(\hat{\theta})\right]\right\} d\theta.$$

Recalling that $h'(\hat{\theta}) = 0$, we have

$$I \approx \int f(\hat{\theta}) \exp\left\{-n\left[h(\hat{\theta}) + \frac{(\theta - \hat{\theta})^2}{2} h''(\hat{\theta})\right]\right\} d\theta$$

$$= f(\hat{\theta}) \exp[-nh(\hat{\theta})] \int \exp\left[\frac{-n(\theta - \hat{\theta})^2}{2\sigma^2}\right] d\theta$$

$$= f(\hat{\theta})\sqrt{\frac{2\pi}{n}} \, \sigma \, \exp[-nh(\hat{\theta})].$$

Intuitively, if $\exp[-nh(\theta)]$ is very peaked about $\hat{\theta}$, then the integral can be well approximated by the behavior of the integrand near $\hat{\theta}$. More formally, it can be shown that

$$I = \hat{I}\left\{1 + O\left(\frac{1}{n}\right)\right\}.$$

To calculate moments of posterior distributions, we need to evaluate expressions such as:

$$E\{g(\theta)\} = \frac{\int g(\theta) \exp\{-nh(\theta)\}d\theta}{\int \exp\{-nh(\theta)\}d\theta}, \qquad (3.2.1)$$

where $\exp\{-nh(\theta)\} = L(\theta \mid Y)p(\theta)$. Tierney and Kadane (1986) provide two approximations to $E\{g(\theta)\}$. The term n refers to the sample size.

Result 3.2.1. $E\{g(\theta)\} = g(\hat{\theta})[1 + O(1/n)]$

To see this, apply Laplace's method to the numerator of (3.2.1) with $f = g$, to obtain the approximation

$$g(\hat{\theta})\sqrt{\frac{2\pi}{n}} \, \sigma \exp[-nh(\hat{\theta})].$$

Next, apply Laplace's method to the denominator of (3.2.1) with $f = 1$, to obtain the approximation

$$\sqrt{\frac{2\pi}{n}} \, \sigma \exp[-nh(\hat{\theta})].$$

Tierney and Kadane (1986) show that the resulting ratio $g(\hat{\theta})$ has error $O(1/n)$. Mosteller and Wallace (1964) present related results.

Result 3.2.2.

$$E\{g(\theta)\} = \frac{\sigma^*}{\sigma} \, \{\exp[-nh^*(\theta^*)]\}/\{\exp[-nh(\hat{\theta})]\}$$
$$\times [1 + O(1/n^2)]$$

To see this, first apply Laplace's method to the numerator of (3.2.1) with $f = 1$, g positive, and $-nh^*(\theta) = -nh(\theta) + \log(g(\theta))$, where θ^* is the mode of $-h^*(\theta)$ and

$$\sigma^* = \left[\frac{\partial^2 h^*}{\partial \theta^2}\Big|_{\theta^*}\right]^{-1/2}.$$

Next, apply Laplace's method to the denominator with $f = 1$. Tierney and Kadane (1986) show that the resulting ratio has error $O(1/n^2)$. Again, Mosteller and Wallace (1964) present related results.

For multivariate θ,

$$\sum{}^* = \left[\frac{\partial^2 h^*}{\partial \theta^2}\Big|_{\theta^*}\right]^{-1}, \quad \sum = \left[\frac{\partial^2 h}{\partial \theta^2}\Big|_{\hat{\theta}}\right]^{-1}$$

and

$$E[g(\theta)] = \left(\frac{\det \sum^*}{\det \sum}\right)^{1/2} \frac{\exp[-nh^*(\theta^*)]}{\exp[-nh(\hat{\theta})]} \left\{1 + O\left(\frac{1}{n^2}\right)\right\}.$$

We will return to these results in Section 5.6 when we discuss the poor man's data augmentation algorithm.

EXAMPLE. *Poisson Data*

Let $y_1, \ldots, y_n$ be an *iid* sample from a Poisson(θ) distribution. It then follows that the likelihood $L(\theta \mid y_1, \ldots, y_n)$ is given by $e^{-n\theta}\theta^{\Sigma y_i}$. Under the Gamma (α, β) prior (proportional to $\theta^{\alpha-1}e^{-\beta\theta}$, where α and β are known), the posterior distribution is gamma with parameters $\tilde{\alpha} = \sum_{i=1}^{n} y_i + \alpha$ and $\tilde{\beta} = \beta + n$. It then follows that the posterior mean is *exactly* $\frac{\tilde{\alpha}}{\tilde{\beta}}$. We now examine the second-order approximation in light of the known exact result.

To compute the second-order approximation to the posterior mean, we note that

$$-nh(\theta) = \log\, L(\theta \mid Y) + \log\, p(\theta)$$

$$= -n\theta + \sum_{i=1}^{n} y_i \log(\theta) + (\alpha - 1)\log(\theta) - \beta\theta$$

$$= (\tilde{\alpha} - 1)\log(\theta) - \theta\tilde{\beta}.$$

It is easy to check that

$$\hat{\theta} = \frac{\tilde{\alpha} - 1}{\tilde{\beta}}$$

and that

$$\sigma = \frac{(\tilde{\alpha} - 1)^{1/2}}{\tilde{\beta}}.$$

Note that

$$-nh^*(\theta) = -nh(\theta) + \log g(\theta) = -nh(\theta) + \log(\theta) = \tilde{\alpha}\log(\theta) - \theta\tilde{\beta}.$$

It follows that

$$\theta^* = \frac{\tilde{\alpha}}{\tilde{\beta}}$$

and

$$\sigma^* = \frac{\tilde{\alpha}^{1/2}}{\tilde{\beta}}.$$

Thus, the second-order approximation is equal to

$$\frac{\frac{\tilde{\alpha}^{1/2}}{\tilde{\beta}}}{\frac{(\tilde{\alpha}-1)^{1/2}}{\tilde{\beta}}} \frac{\left(\frac{\tilde{\alpha}}{\tilde{\beta}}\right)^{\tilde{\alpha}}}{\left(\frac{\tilde{\alpha}-1}{\tilde{\beta}}\right)^{\tilde{\alpha}-1}} \frac{\exp(-\tilde{\alpha})}{\exp(-(\tilde{\alpha}-1))} = \left(\frac{\tilde{\alpha}}{\tilde{\beta}}\right)\left(\frac{\tilde{\alpha}}{\tilde{\alpha}-1}\right)^{\tilde{\alpha}-1/2} e^{-1}.$$

Note that the relative error (relative to the exact posterior mean $\tilde{\alpha}/\tilde{\beta}$) is

$$\left(\frac{\tilde{\alpha}}{\tilde{\alpha}-1}\right)^{\tilde{\alpha}-1/2} e^{-1}.$$

The following table presents the relative error for several values of $\tilde{\alpha}$ (Press, 1989). When $\tilde{\alpha}$ equals 8, the relative error is .15%.

$\tilde{\alpha}$	2	4	6	8
Relative Error	1.0405	1.0069	1.0028	1.0015

3.2.2. Marginalization

Partition the $d \times 1$ vector θ as (θ_1, θ_2), where θ_1 is a scalar and θ_2 is $(d - 1) \times 1$. Let $\hat{\theta}$ maximize the posterior $p(\theta \mid Y)$. For a given value of θ_1, let $\hat{\theta}_2^* = \hat{\theta}_2^*(\theta_1)$ maximize $p(\theta_1, \theta_2 \mid Y)$ with θ_1 fixed, and let

$$\hat{\Sigma}^* = \left[\frac{-\partial^2 \log p(\theta_1, \theta_2 \mid Y)}{\partial \theta_2^2} \bigg|_{\hat{\theta}_2^*} \right]^{-1},$$

which is a $(d - 1) \times (d - 1)$ matrix. Tierney and Kadane (1986) approximate $p(\theta_1 \mid Y)$ with

$$\hat{p}(\theta_1 \mid Y) \propto (\det \hat{\Sigma}^*)^{1/2} p(\theta_1, \hat{\theta}_2^* \mid Y) = (\det \hat{\Sigma}^*)^{1/2} \text{ (profile posterior of } \theta_1).$$

To obtain the marginal of an arbitrary function $g(\theta)$, Tierney, Kass and Kadane (1989) approximate $p(\gamma \mid Y)$ with

$$\hat{p}(\gamma \mid Y) \propto \left[\frac{\det[\Sigma(\gamma)]}{\det[(D_g)^T \Sigma(\gamma)(D_g)]} \right]^{1/2} p[\hat{\theta}(\gamma) \mid Y],$$

where $\hat{\theta}(\gamma)$ maximizes the posterior $p(\theta \mid Y)$ subject to the constraint that $g(\theta) = \gamma$, $\Sigma(\gamma)$ is the inverse Hessian of the log posterior evaluated at $\hat{\theta}(\gamma)$ and D_g is the gradient or Jacobian of g evaluated at $\hat{\theta}(\gamma)$. This approximation may be poor in small samples. See Leonard, Hsu and Tsui (1989), as well as Wong and Li (1992) for alternative approximations.

EXAMPLE. *Stanford Heart Transplant Data*
Turnbull, Brown and Hu (1974) proposed a Pareto model for data from the Stanford Heart Transplant Program. Each patient has an exponential lifetime (i.e. constant hazard) upon entering the program, with the hazard rate varying from patient to patient according to a gamma distribution with parameters p and λ. In this way, the marginal lifetime density is Pareto. A transplant, if it occurs, multiplies the patient's hazard rate by a factor τ, resulting in the likelihood

$$\prod_{i=1}^{n} \frac{p\lambda^p}{(\lambda + x_i)^{p+1}} \prod_{i=n+1}^{N} \left(\frac{\lambda}{\lambda + x_i}\right)^p \prod_{j=1}^{m} \frac{\tau p\lambda^p}{(\lambda + y_j + \tau z_j)^{p+1}} \times$$

$$\prod_{j=m+1}^{M} \left(\frac{\lambda}{\lambda + y_j + \tau z_j}\right)^p,$$

where x_i is the survival time in days for $N = 30$ nontransplant patients, of which $n = 26$ died and 4 were censored, and y_j, z_j are the time to transplant and survival time (beyond transplant), respectively, for the transplant patient ($M = 52$) of whom ($m = 34$) died and 18 were censored.

Figure 3.3 presents the marginal density of p. The solid line is based on the Laplace approximation, while the dashed line is the asymptotic normal approximation.

Superimposed on the solid line is the marginal density obtained by numerical integration (Gauss–Hermite quadrature, see Naylor and Smith, 1982). There is a noticeable discrepancy between the normal approximation and other approaches.

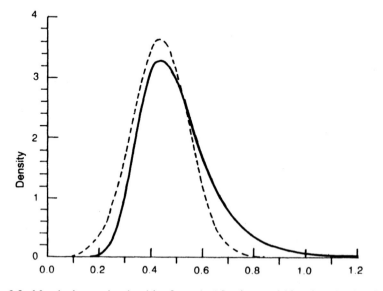

FIGURE 3.3. Marginal posterior densities for p. (—) Laplace and 20-point adaptive Gauss–Hermite approximations; ($\cdots$) asymptotic normal approximation (Tierney and Kadane 1986, Figure 1).

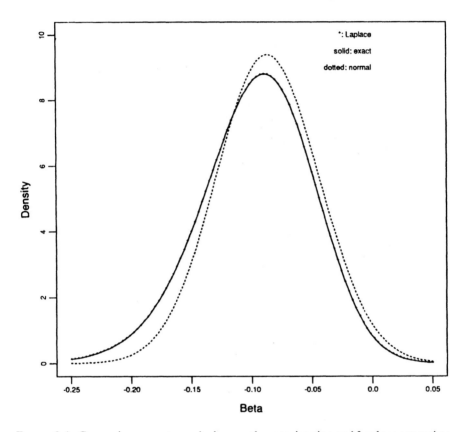

FIGURE 3.4. Comparison: exact marginal, normal approximation and Laplace approximation.

Note that the marginal is somewhat skewed, even though the total sample size is moderate.

EXAMPLE. *Logistic Regression (Continued)*
Figure 3.4 presents the β marginal (starred line) of the likelihood obtained via the Tierney and Kadane (1986) marginalization procedure. The solid line represents the marginal of the likelihood obtained via numerical integration in Section 3.1. The dotted line is the normal approximation. There is a slight discrepancy between the normal approximation and the other approaches. However, the Laplace approach and the numerically integrated marginal are quite congruent. The Tierney and Kadane (1986) marginal for β was obtained by calculating the profile likelihood for β (weighted by $\hat{\Sigma}^*$) on a grid of points between -0.25 and 0.05. The curve was

then normalized to integrate to unity by calculating the area under this curve and then dividing through by this constant. The trapezoidal rule was used to calculate the normalizing constant.

Figure 3.5 presents the $-\alpha/\beta$ marginal. The starred line was obtained by applying the Tierney and Kadane (1986) marginalization procedure to the likelihood parameterized in terms of α and ρ, where $\rho = -\alpha/\beta$, with a flat prior on ρ. The solid line represents the marginal obtained via numerical integration with a flat prior on ρ. The normal approximation (δ-method) is given by the dotted line. The Laplace approach and numerical integration yield virtually identical curves in this example.

It is noted that in this case we have applied the Tierney and Kadane (1986) marginalization procedure to the reparameterized likelihood $L(\alpha, \rho)$ supposing a flat prior on ρ, thus leading to an improper posterior (the improper posterior in the

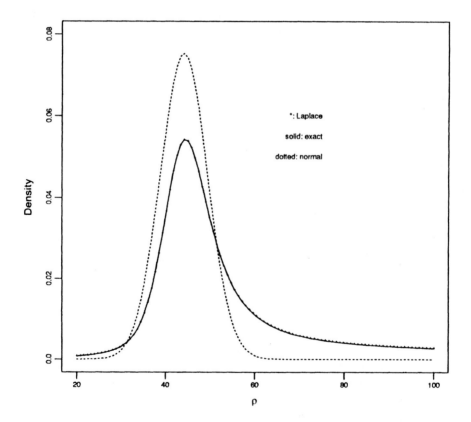

FIGURE 3.5. Comparison: exact marginal, normal approximation, Laplace approximation.

above figure was truncated at $\rho = 100$). The Tierney, Kass and Kadane (1989) approach would assign the prior $| \alpha | / \rho^2$, the Jacobian of the transformation, to ρ and would result in a more symmetric curve, which is closer to the delta method (normal) solution.

3.3 Monte Carlo Methods

3.3.1. Monte Carlo

Consider the problem of approximating the integral

$$J(y) = \int f(y \mid x)g(x)dx = E_g[f(y \mid x)] < +\infty,$$

where y and x may be vectors. If $g(x)$ is a density, then the Monte Carlo method (after the famous casino in Monaco) approximates $J(y)$ as

$$\hat{J}(y) = \frac{1}{n} \sum_{i=1}^{n} f(y \mid x_i),$$

where $x_1, \ldots, x_n \overset{\text{iid}}{\sim} g(x)$. By the Law of Large Numbers,

$$\hat{J}(y) \overset{\text{a.s.}}{\to} J(y)$$

(Geweke, 1989). The estimated Monte Carlo standard error of $\hat{J}(y)$ is

$$\frac{1}{\sqrt{n}} \sqrt{\frac{\sum_{i=1}^{n}[f(y \mid x_i) - \hat{J}(y)]^2}{n-1}}.$$

Extensive use will be made of the method of Monte Carlo when we consider the data augmentation algorithm (Chapter 5).

EXAMPLE. *Area of the Quarter Circle*
Let $f(y \mid x) = \sqrt{1 - x^2}$ and let $g(x)$ be the density of the uniform distribution on $[0,1]$, i.e. we wish to compute one-quarter of the area of the unit circle ($\pi/4 = 0.7854\ldots$)

Five thousand pseudo-random uniform deviates were generated using the RANDOM command in MINITAB. $\hat{J}$ was then taken to be

$$\hat{J} = \frac{1}{5000} \sum_{i=1}^{5000} \sqrt{1 - x_i^2} = 0.7865.$$

The standard error of $\hat{J}$ is

$$\sqrt{\frac{2/3 - \left(\frac{\pi}{4}\right)^2}{5000}} = 0.0032.$$

There are a number of techniques that can improve the precision of the Monte Carlo estimate. A general discussion of variance-reduction techniques, such as antithetic variates and control variates, is given in Rubenstein (1981).

3.3.2. Composition

Suppose $f(y \mid x)$ is a density where x and y may be vectors. To obtain a sample $y_1, \ldots, y_m \overset{iid}{\sim} J(y) = \int f(y \mid x)g(x)dx$, one may use the method of composition:

1. Draw $x^* \sim g(x)$.
2. Draw $y^* \sim f(y \mid x^*)$.

Steps 1 and 2 are repeated m times. The pairs $(x_1, y_1), \ldots, (x_m, y_m)$ are an *iid* sample from the joint density $h(x, y) = f(y \mid x)g(x)$, while the quantities $y_1, \ldots, y_m$ are an *iid* sample from the marginal $J(y)$.

When x is a discrete random variable, taking on values $0, 1, 2, \ldots$, select an integer (i) with probability $g(i)$ and draw y^* from $f_i(y)$. The resulting deviate is an observation from the distribution with density

$$\sum_{j=1}^{\infty} f_j(y)g(j).$$

When x takes on a finite number of values, n, select an integer (i) with probability $1/n$, draw y^* from $f_i(y)$ and then assign mass $g(i)$ to y^*. The resulting deviate (with assigned mass $g(i)$) is drawn from the distribution with density

$$\sum_{j=1}^{n} f_j(y)g(j).$$

For a larger number of components, consider

$$J(y) = \int f(y \mid x, s, t)g(x \mid s, t)h(s \mid t)\, j(t)dx\, ds\, dt,$$

where $j(.), h(. \mid .), g(. \mid .)$ and $f(. \mid .)$ are densities. To sample from $J(y)$:

1. Draw t^* from $j(t)$.
2. Draw s^* from $h(s \mid t^*)$.
3. Draw x^* from $g(x \mid s^*, t^*)$.
4. Draw y^* from $f(y \mid x^*, s^*, t^*)$.

The 4-tuple (y^*, x^*, s^*, t^*) is an observation from the joint distribution, while y^* is an observation from the marginal $J(y)$. The method of composition is a key technique underlying the data augmentation algorithm of Chapter 5.

EXAMPLE. *Predictive Distribution*

Let $Y = (y_1, \ldots, y_n)$ be the observed data and let y_f denote a (univariate) future observation. Then the predictive distribution is given by

$$p(y_f \mid Y) = \int p(y_f \mid Y, \theta) p(\theta \mid Y) d\theta.$$

To compute an approximation to $p(y_f \mid Y)$ via the Monte Carlo method:

1. Draw a sample $\theta_1, \ldots, \theta_m \overset{iid}{\sim} p(\theta \mid Y)$.
2. Let

$$\hat{p}(y_f \mid Y) = \frac{1}{m} \sum_{i=1}^{m} p(y_f \mid Y, \theta_i).$$

To sample from $p(y_f \mid Y)$, apply the method of composition:

1. Draw θ^* from $p(\theta \mid Y)$.
2. Draw y^* from $p(y_f \mid Y, \theta^*)$.

Steps 1 and 2 are repeated m times to yield the desired sample. This sample may then be used to compute the posterior distribution of any functional of the future observation.

As an example of a predictive distribution, consider the linear model

$$y = x\theta + \varepsilon,$$

where y is an $n \times 1$ vector of associations, x is an $n \times d$ matrix of known constants, θ is a $d \times 1$ vector of unknown regression coefficients, and ε is an $n \times 1$ vector of errors, where $\varepsilon_i \sim N(0, \sigma^2)$ and σ is unknown. As noted in Section 2.1, under the prior $p(\theta, \sigma^2) \propto \sigma^{-2}$,

$$p(\theta, \sigma^2 \mid Y) = p(\sigma^2 \mid s^2) p(\theta \mid \hat{\theta}, \sigma^2),$$

where the marginal distribution of σ^2 is given by $vs^2 \chi_v^{-2}$ ($v = n - d$ and $s^2 = (y - \hat{y})(y - \hat{y})/v$). The conditional distribution of θ given σ^2 is

$$N[\hat{\theta}, \sigma^2(x^T x)^{-1}],$$

where $\hat{y} = x\hat{\theta}$ and $\hat{\theta} = (x^T x)^{-1} x^T y$.

Note that in this case, $p(y_f \mid Y, \sigma^2, \theta)$ is the normal distribution with mean $x_f \theta$ and variance σ^2. We can approximate

$$p(y_f \mid Y) = \int p(y_f \mid Y, \sigma^2, \theta) p(\theta, \sigma^2 \mid Y) d\theta \, d\sigma^2$$

via the Monte Carlo method. First draw an observation from $p(\theta, \sigma^2 \mid Y)$:

1a. Draw σ_*^2 from $p(\sigma^2 \mid s^2)$.
1b. Draw θ^* from $p(\theta \mid \hat{\theta}, \sigma_*^2)$.

Steps (1*a*) and (1*b*) are repeated to obtain a sample $(\theta_1, \sigma_1^2), \ldots, (\theta_m, \sigma_m^2)$ from $p(\theta, \sigma^2 \mid Y)$. The approximation $\hat{p}(y_f \mid Y)$ is then given by

$$\frac{1}{m} \sum_{i=1}^{m} \phi(y_f \mid x_f \theta_i, \sigma_i^2),$$

where $\phi(x \mid a, b)$ is the (univariate) normal density function with mean a and variance b, i.e. $\exp[-\frac{1}{2}(x-a)^2/b]/\sqrt{2\pi b}$.

To draw σ_*^2 from $p(\sigma^2 \mid s^2)$, we draw a chi-square deviate on $n - d$ degrees of freedom, form the reciprocal of this number and then multiply by $(n - d)s^2$. To draw θ^* from $p(\theta \mid \hat{\theta}, \sigma_*^2)$, we draw a deviate from the multivariate normal distribution $N[\hat{\theta}, \sigma_*^2(x^T x)^{-1}]$. Thus, the tuple (σ_*^2, θ^*) is an observation drawn from $p(\theta, \sigma^2 \mid Y)$.

To draw a sample from $p(y_f \mid Y)$, apply the method of composition:

1a. Draw σ_*^2 from $p(\sigma^2 \mid s^2)$.
1b. Draw θ^* from $p(\theta \mid \hat{\theta}, \sigma_*^2)$.
 2. Draw y_f from $N(x_f \theta^*, \sigma_*^2)$.

Steps (1*a*), (1*b*) and (2) are repeated to obtain the required sample.

Thompson and Miller (1986) use the method of composition to sample future paths from a time series.

3.3.3. *Importance Sampling and Rejection/Acceptance*

Consider the problem of calculating the integral

$$J(y) = \int f(y \mid x)g(x)dx.$$

Previously, it was assumed that one can sample directly from $g(x)$. Importance sampling is of use when one cannot directly sample from $g(x)$. Let $I(x)$ be a density that is easy to sample from and approximates $g(x)$. The method of importance sampling approximates $J(y)$ as

1. Draw $x_1, \ldots, x_m \overset{iid}{\sim} I(x)$.
2. $\tilde{J}(y) = \sum_{i=1}^{m} w_i f(y \mid x_i)/\sum_{i=1}^{m} w_i$,

where $w_i = g(x_i)/I(x_i)$.

Importance sampling gives more weight to regions where $I(x) < g(x)$ and downweights regions where $I(x) > g(x)$ to correctly calculate $J(y)$, given a sample from $I(x)$.

Geweke (1989) has shown that if the support of $I(x)$ includes the support of $g(x)$, the x_i's are an *iid* sample from $I(x)$, and $J(y)$ exists and is finite, then

$$\tilde{J}(y) \overset{a.s.}{\to} J(y).$$

The first condition is sensible, for if the support of $I(x)$ is strictly contained in the support of $g(x)$, then there would be no hope of generating deviates in the complement of the support of $I(x)$. The rate of convergence depends on how closely $I(x)$ mimics $g(x)$. As noted by Geweke (1989), it is important that the tails of $I(x)$ do not decay faster than the tails of $g(x)$.

Furthermore, it can be shown (Geweke, 1989) that the Monte Carlo standard error of $\bar{J}(y)$ is estimated by

$$\sqrt{\sum_{i=1}^{m}\{f(y \mid x_i) - \bar{J}(y)\}^2 w_i^2 / \sum_{i=1}^{m} w_i}.$$

In this way, the standard error of $\bar{J}(y)$ is inflated (the effective Monte Carlo sample size is decreased) if $I(x)$ poorly approximates $g(x)$.

One can apply the method of composition to the integral

$$J(y) = \int f(y \mid x)\left[\frac{g(x)}{I(x)}\right] I(x)dx$$

to obtain a sample from the marginal of y. Note that

1. draw x^* from $I(x)$
2. draw y^* from $f(y \mid x^*)$

yields an observation from the joint density $f(y \mid x)I(x)$. Assigning mass $w_i / \sum_{j=1}^{m} w_j$ to each of the $y_1, \ldots, y_m$, where $w_i = g(x_i)/I(x_i)$, yields a weighted sample from an approximation to $J(y)$. One can carry along these weights in all subsequent analyses of $y_1, \ldots, y_m$. Alternatively, one can draw $y_1^*, \ldots, y_n^*$ from the discrete distribution over $(y_1, \ldots, y_m)$ placing mass w_i on y_i to obtain an *iid* sample. Crucial to this method is the necessity that the support of $g(x)$ is contained in the support of $I(x)$. Smith and Gelfand (1992) call this sample a "weighted bootstrap" sample from an approximation to $J(y)$. They prove that the approximation improves as m increases. Rubin (1987a) uses this idea as part of the SIR algorithm (see Section 5.7). In this regard, note that this sampling approach only requires that $g(x)$ be known up to a proportionality constant, since the constant cancels out in $w_i / \sum_{i=1}^{m} w_i$. This observation is key, since we have avoided the need to perform the integration to standardize $g(x)$. Note, however, that $\sum_{i=1}^{m} w_i/m$ is a consistent estimator of $\int g(x)dx$.

In the situation where there is a finite known constant $M > 0$ such that $g(x)/I(x) \leq M$, for all x, one can realize an *iid* sample from $J(y)$ exactly. The algorithm is given as:

1. Generate x^* from $I(x)$.
2. Generate u from the Uniform [0, 1] distribution, independent of x^*.
3. If $u \leq g(x^*)/[M \cdot I(x^*)]$, then accept x^*; otherwise, repeat steps 1–3.
4. Draw y^* from $f(y \mid x^*)$.

Steps (1)–(3) above form a *rejection/acceptance algorithm*. The function $M \times I(x)$ is called the majorizing function. It is easy to see that x^* is an observation from $g(x)$. Note that

$$P(X < x^*) = P\left(X < x^* \mid U < \frac{g(X)}{MI(X)}\right)$$
$$= \frac{\int_{-\infty}^{x^*} \int_0^{g(x)/MI(x)} I(x)\, du\, dx}{\int_{-\infty}^{\infty} \int_0^{g(x)/MI(x)} I(x)\, du\, dx}$$
$$= \frac{\int_{-\infty}^{x^*} g(x)\, dx}{\int_{-\infty}^{\infty} g(x)\, dx}$$

Differentiation with respect to x^* yields $g(.)$ as the density.

Rejection/acceptance allows one to indirectly sample from $g(x)$, if the functional form of $g(x)$ is known and a sample from an approximating distribution $I(x)$ is given. Note that the probability of accepting x^* is equal to $\int g(x)dx/M$. Hence, if the approximation of $I(x)$ to $g(x)$ is poor in some regions (forcing $\int g(x)dx/M$ to be small), then the algorithm may have a low acceptance rate.

EXAMPLE. *Statistical Uses of Importance Sampling*
Suppose that $\theta_1, \ldots, \theta_m$ are an *iid* sample from the posterior $p(\theta \mid Y)$ calculated under the prior $p(\theta)$. Assigning mass proportional to the weight $w = q(\theta)/p(\theta)$ to the sampled values realizes an *iid* sample from the posterior $p'(\theta \mid Y)$ calculated under the prior $q(\theta)$. Thus, one may wish to work with a convenient prior to obtain the sample $\theta_1, \ldots, \theta_m$ and then weight these observations to realize a sample from the posterior of interest. However, since the Monte Carlo standard error of some function of the θ_i's will depend on the weights, a high variance of the weights will have the effect of decreasing the effective Monte Carlo sample size. Thus, for the method to be effective, one must find a trial density that is easy to sample from yet leads to reasonably distributed weights.

To judge the sensitivity of the posterior to an observation y_i, one may wish to delete the corresponding term from the likelihood and examine the resulting posterior. Having drawn $\theta_1, \ldots, \theta_m$ from $p(\theta \mid Y)$, a sample from $p(\theta \mid Y_{-i})$, i.e. the posterior based on the reduced sample, is obtained by weighting θ_l proportional to $1/L(\theta_l \mid y_i)$, where $L(\theta_l \mid y_i)$ is the contribution to the likelihood corresponding to y_i evaluated at θ_l. As noted above, the effective Monte Carlo sample size will be reduced if the terms $1/L(\phi \mid y_i)$, as a function of ϕ, are highly variable. Such a situation, however, would suggest that the posterior is sensitive to y_i. Tanner (1991) shows how to use sampling based algorithms for prior sensitivity and for case influence analyses. See also Kass, Tierney and Kadane (1989), Smith and Gelfand (1992) and Kong, Liu and Wong (1994). The key point is that the computation does not have to be redone from scratch for each change in the likelihood or prior.

EXAMPLE. *Logit Model (Continued)*
To illustrate importance sampling, approximate the posterior (under a flat prior) of the logit model by a normal distribution with mean $\hat{\theta}$ (MLE) and variance–covariance matrix $c \sum$, when $\sum$ is the inverse information at $\hat{\theta}$ and

c is a constant. Let $I_c(\alpha, \beta) = I_c(\theta)$ denote the density of the bivariate normal distribution, let $L(\alpha, \beta \mid Y) = L(\theta \mid Y)$ denote the bivariate likelihood and let the weight $w_i = L(\theta_i \mid Y)/I_c(\theta_i)$, where θ_i is a value of θ.

Trial runs ($m = 100$) were conducted by sampling from the normal distribution with variance–covariance matrices $\sum$, $1.5 \sum$ and $2.0 \sum$. It was found that the coefficient of variation of the weights w_i, ($i = 1, \ldots, 100$) was minimized from the covariance matrix $1.5 \sum$.

A sample of size 10,000 was then drawn from this bivariate normal distribution. The mass $w_i / \sum_{j=1}^{10,000} w_j$ was assigned to the bivariate normal realization θ_i, where $w_i = L(\theta_i \mid Y)/I(\theta_i)$.

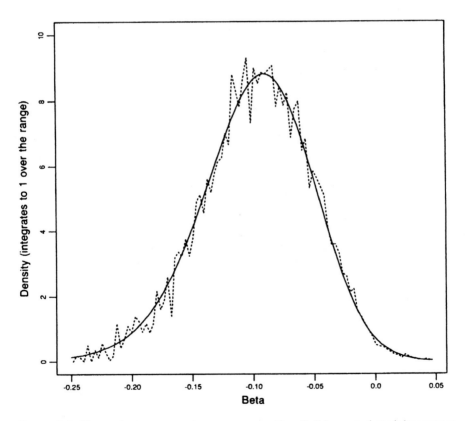

FIGURE 3.6. Comparison: exact vs. importance sampling. Solid: exact; dotted: importance sampling.

The resulting β marginal (dashed line) is presented in Figure 3.6, along with the marginal obtained via numerical integration (solid line). As can be seen, the importance sampling approach follows the marginal. Using a smaller number of bins or adopting a kernel estimator would further smooth the estimate. Alternatively, if the normalized version of $L(\beta \mid \alpha, Y)$ is available, one can use the Rao–Blackwell estimator (see Section 6.2.1)

$$\frac{\sum_{j=1}^{10.000} w_j \, L(\beta \mid \alpha_j, Y)}{\sum_{j=1}^{10.000} w_j},$$

where the normalized conditionals are used in this mixture.

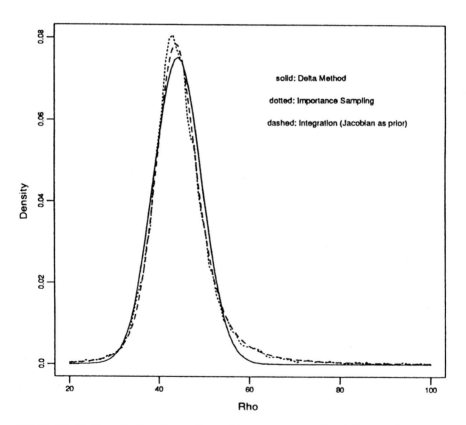

FIGURE 3.7. Delta method vs. integration and importance sampling (all curves integrate to 1 over the displayed range).

The posterior of $-\alpha/\beta$ is easily obtained by computing $-\alpha/\beta$ for all the simulated bivariate normal deviates and assigning mass $w_i / \sum_{j=1}^{10.000} w_j$ to $-\alpha_i/\beta_i$. This estimate of the marginal is presented in Figure 3.7 (dotted line) along with the δ-method approximation (solid line). The dashed line represents the marginal obtained via numerical integration (Jacobian prior). The importance sampling result is quite congruent with the result obtained via numerical integration under the Jacobian prior.

As noted in Section 3.1, by reparameterizing the likelihood in terms of α and $\rho = -\alpha/\beta$, i.e. $L(\alpha, -\alpha/\rho \mid Y)$, and assuming a flat prior on $-\alpha/\beta$, one obtains the nonproper marginal. However, if the flat prior is put on (α, β), leading to the prior $\beta^2/ \mid \alpha \mid$ on ρ, i.e. the Jacobian of the transformation, the more symmetric marginal is obtained.

Sampling from an approximation $I(\alpha, \beta)$ to $L(\alpha, \beta \mid Y)$ and computing the weights

$$\frac{L(\alpha_i, \beta_i \mid Y)}{I(\alpha_i, \beta_i)} = \frac{L(\alpha_i, -\alpha_i/\rho_i \mid Y) \mid \alpha \mid /\rho_i^2}{I(\alpha_i, -\alpha_i/\rho_i) \mid \alpha \mid /\rho_i^2},$$

where $\rho_i = -\alpha_i/\beta_i$, yields the proper marginal due to the cancellation of the Jacobian. The nonproper marginal can be obtained by multiplying w_i by $\rho_i^2/ \mid \alpha_i \mid$. However, in this case it seems more appropriate to put the flat prior on (α, β) rather than on ρ. Posterior distributions under other priors may be obtained by multiplying w_i by the appropriate term.

Exercises

1. Consider two urns each containing an unknown mixture of blue and white marbles. A random sample of size 18 (with replacement) is drawn from urn #1 and a random sample of size 6 (with replacement) is drawn from urn #2. (The numbers 18 and 6 are fixed by design.) Of the 18 selected marbles from urn #1, 14 are blue. The corresponding number of blue marbles from urn #2 is 2.

   a. Let π denote the proportion of blue marbles in urn #1 and let ψ denote the corresponding proportion in urn #2. Under the (i) Haldane, (ii) flat and (iii) noninformative priors, compute

   $$p\left(\log\left(\frac{\pi}{1-\pi}\right) > \log\left(\frac{\psi}{1-\psi}\right) \mid \text{data}\right)$$

   using the normal approximation found in Section 2.4.

   b. Repeat (a) by drawing deviates from the appropriate beta distributions. Quantify the Monte Carlo error in your value.

   c. Compare your results in (a) and (b) to the p-value obtained via Fisher's exact test.

2. Recalling the genetic linkage model of Exercise 9 in Chapter 2:

a. Use the second-order Laplace approximation to compute the posterior mean and posterior standard deviation for the data $Y = (125, 18, 20, 34)$.

b. Repeat (a) for the data $Y = (14, 0, 1, 5)$.

3. For the genetic linkage model:

a. Use importance sampling to obtain the posterior mean and posterior standard deviation for the data $Y = (125, 18, 20, 34)$. Use the matching normal distribution as the importance function. Compare your importance sampling estimates of the posterior mean and posterior standard deviation to those obtained via Laplace's method. Draw the histogram of the weights and compute their standard deviation.

b. Repeat (a) for the data $Y = (14, 0, 1, 5)$. Compare the histograms of the weights for both data sets. Discuss the adequacy of the importance sampling estimate for each data set.

c. Repeat (a) and (b) using a matching beta distribution as the importance function.

d. Use importance sampling to obtain the normalized likelihood for the data $Y = (125, 18, 20, 34)$. Plot this curve and the true normalized likelihood in the same figure.

e. Repeat (d) for the data $Y = (14, 0, 1, 5)$. Plot this curve and the true normalized likelihood in the same figure.

4.a. Solve the following problem posed by the Reverend Thomas Bayes in his essay "Essay Towards Solving a Problem in the Doctrine of Chances," which was published in the *Philosophical Transactions of the Royal Society* (London) in 1763:

> *Given* the number of times in which an unknown event has happened and failed: *Required* the chance that the probability of its happening in a single trial lies somewhere between any two degrees of probability that can be named.

In other words, if the number of successful happenings of the event is p and the failures q, and if the named "degrees" of probability are b and f, respectively, compute:

$$\int_b^f x^p (1 - x)^q dx / \int_0^1 x^p (1 - x)^q dx$$

via importance sampling. Take $p = 1$, $q = 4$, $b = 0.7$, and $f = 0.9$. Quantify the Monte Carlo error in your value.

b. Repeat the calculation using numerical integration. Compare the results of (a) and (b).

c. Suppose that 241,945 girls and 251,527 boys were born in Paris between 1745 and 1770. What is the posterior probability that p (the probability of a female birth) is greater than 0.5? Compare your result to that obtained using a normal approximation to the posterior.

5. In this exercise, we reanalyze the Darwin data in Exercise 2 in Chapter 2 using a t distribution. In a Bayesian analysis, we can reduce the effect of an

outlier by replacing the normal distribution with a distribution with thicker tails. In this context, an outlying observation has the interpretation of arising from a long-tailed distribution rather than from a normal distribution with a large variance. By varying the degrees of freedom, one can also examine the sensitivity of the analysis to the normality assumption.

a. Suppose that the data follow the t distribution with mean μ, variance σ^2 (unknown), on 4 degrees of freedom. Under the noninformative prior, use numerical integration to obtain the posterior marginal of μ. What is your best guess for μ? Obtain a 95% credible interval for μ and interpret this interval. Compute $p(\mu > 0 \mid Y)$.

b. Repeat (a) using importance sampling. Compare the results of (a) and (b).

c. To assess the adequacy of a model, Gelman et al. (1995) suggest that one compare the observed data set to samples from the predictive distribution $p(y_f \mid Y)$. Draw repeated samples of size 15 from $p(y_f \mid Y)$ using the present model. For each sample, draw a quantile-quantile plot of the simulated data versus the observed data. An inadequate model would result in departures from linearity in these plots.

d. Repeat (c) for t distributions with 8, 16, 32 and ∞ degrees of freedom. In this class of models, is there a model that is most appropriate for the Darwin data?

6. a. Under the likelihood $\theta^x (1-\theta)^{n-x}$ and the Beta(a, b) prior (a and b known) compute the exact posterior mean. Repeat the calculation using the second-order Laplace approximation. Evaluate the relative error for the data $n = 5, x = 3$, and the prior values $a = b = 1/2$. What is the relative error when $n = 25, x = 15$ (same prior)?

b. Transform the problem to the $\sin^{-1} \sqrt{\theta}$ scale and repeat (a). Compare the results of (a) and (b).

c. Repeat (a) and (b) for the posterior variance.

7. a. Under the likelihood $\theta^y (1-\theta)^{n-y}$ and the Beta(a, b) prior (a and b known), what is the posterior distribution?

b. Given that in a future experiment of N trials, the probability of x successes is $\binom{N}{x}\theta^x (1-\theta)^{N-x}$, derive the predictive distribution. The resulting probability density function is known as the *Beta-Binomial distribution*.

c. Suppose we have observed y successes in the first n trials. What is the probability that the next trial is a success? The case where $y = n$ is known as Laplace's Rule of Succession.

8. The following data represent the results of 71 experiments on F344 female rats (Tarone, 1982). The quantity (y_i) represents the number of rats who developed an endometrial stromal polyp out of the total number of rats in the particular experiment (n_i). For example, there were seven experiments where none of the rats (out of 20) developed a polyp.

Control Rates for Endometrial Stromal Polyp from 71 Experiments
with Female F344 Rats (Tarone, 1982)

y_j/n_j	Frequency	y_j/n_j	Frequency	y_j/n_j	Frequency
0/20	7	5/49	1	5/22	1
0/19	4	2/19	1	11/46	1
0/18	2	5/46	1	12/49	1
0/17	1	2/17	1	5/20	2
1/20	4	7/49	1	6/23	1
1/19	2	7/47	1	5/19	1
1/18	2	3/20	2	6/22	1
2/27	1	2/13	1	6/20	3
2/25	1	9/48	1	16/52	1
2/24	1	10/50	1	15/47	1
2/23	1	4/20	7	15/46	1
2/20	6	10/48	1	9/24	1
1/10	1	4/19	3	4/14	1

We will model these data by assuming that the y_i, $j = 1, \ldots, 71$, are independent binomial random variables with parameters n_j and θ_j, where n_j is known and the θ_j are an *iid* sample from the beta distribution with parameters α and β. To complete the specification of the model, we suppose a flat prior on [0,1000] for α and β. This model assumes that the experiments are similar in factors known to affect the magnitude of the tumor rate, e.g. length of time on study, housing conditions, type of food, method of feeding, and possibly animal supplier (see Tarone, 1982). Let $\theta = (\theta_1, \ldots, \theta_{71})$ and $\Gamma(.)$ be the gamma function.

a. Show that the posterior $p(\theta, \alpha, \beta \mid Y)$ is proportional to $p(\theta \mid \alpha, \beta) \times p(Y \mid \alpha, \beta, \theta)$, for α and β in [0,1000] and zero elsewhere.

b. Show that $p(\theta \mid \alpha, \beta) p(Y \mid \alpha, \beta, \theta) \propto$

$$\prod_{j=1}^{71} \frac{\Gamma(\alpha + \beta)}{\Gamma(\alpha)\Gamma(\beta)} \theta_j^{\alpha-1}(1 - \theta_j)^{\beta-1} \prod_{j=1}^{71} \theta_j^{y_j}(1 - \theta_j)^{n_j - y_j},$$

for α and β in [0,1000] and zero elsewhere.

c. Show that $p(\theta \mid \alpha, \beta, Y)$ is equal to

$$\prod_{j=1}^{71} \frac{\Gamma(\alpha + \beta + n_j)}{\Gamma(\alpha + y_j)\Gamma(\beta + n_j - y_j)} \theta_j^{\alpha+y_j-1}(1 - \theta_j)^{\beta+n_j-y_j-1},$$

i.e. the θ_j's have (conditionally) independent beta densities.

d. Show (Gelman et al. 1995) that $p(\alpha, \beta \mid Y)$ is proportional to

$$f(\alpha, \beta) = \prod_{j=1}^{71} \frac{\Gamma(\alpha + \beta)}{\Gamma(\alpha)\Gamma(\beta)} \frac{\Gamma(\alpha + y_j)\Gamma(\beta + n_j - y_j)}{\Gamma(\alpha + \beta + n_j)}.$$

e. Draw a contour plot of $f(\alpha, \beta)$. Does this plot suggest a good importance function? Now transform the variables to $\left(\log\left(\frac{\alpha}{\beta}\right), \log(\alpha + \beta)\right)$ and redo the plot. Use importance sampling to draw a sample of size 10,000 from $p(\alpha, \beta \mid y)$. (*Hint:* Transform to $\left(\log\left(\frac{\alpha}{\beta}\right), \log(\alpha + \beta)\right)$. Now use importance sampling to draw from this distribution and then transform back to the (α, β) scale.)

f. Use the sample you obtained in (e) and the method of composition to obtain a sample from the posterior distribution for θ_{71}, i.e. the θ_j corresponding to the observed rate of 4/14. What is the marginal posterior mean and marginal posterior standard deviation for θ_{71}? Compare this distribution to the observed rate of 4/14. More generally, is there shrinkage from the observed rate of y_j/n_j toward the "grand mean" of $262/1725 = 0.152$?

4

The *EM* Algorithm

In the previous chapters, we examined various methods that are applied directly to the likelihood or to the posterior density. In this and the following chapters, we examine the data augmentation algorithms, including the *EM* algorithm and the data augmentation algorithm. All of these data augmentation algorithms share a common approach to problems: rather than performing a complicated maximization or simulation, one augments the observed data with "stuff" (latent data) that simplifies the calculation and subsequently performs a series of simple maximizations or simulations. This "stuff" can be the "missing" data or parameter values. The principle of data augmentation can then be stated as follows: Augment the observed data Y with latent data Z so that the augmented posterior distribution $p(\theta \mid Y, Z)$ is "simple." Make use of this simplicity in maximizing/marginalizing/calculating/sampling the observed posterior $p(\theta \mid Y)$.

4.1 Introduction

To motivate the *EM* algorithm, consider the situation depicted in Figure 4.1. The x's in the plot represent failure event times, while the o's represent right-censored event times. Given a current guess to the slope and intercept, we can impute a value for each of the censored event times. In particular, assuming normally distributed errors, this imputation would be based on the conditional normal distribution, conditional on the fact that the (unobserved) failure event time is larger than the (observed) censored time. Having "filled in" the "holes", update the estimates of the slope and intercept and iterate to obtain the parameter estimates. Thus, a complicated analysis is replaced by a series of simpler analyses. This situation is treated in detail later in this section.

This idea of filling in missing values with estimated values and then updating parameter estimates is an *EM* algorithm in the situation where the log-augmented posterior is linear in the latent data Z. Estimates can be severely biased when this simple approach is applied in general (Little and Rubin, 1983). In more general situations, expected sufficient statistics, rather than individual observa-

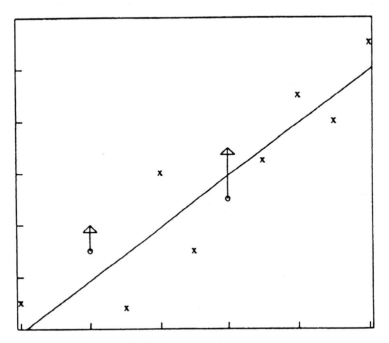

FIGURE 4.1. Right-censored event time data.

tions, are computed at each iteration. In the most general situation, the expected log-augmented posterior is computed at each iteration.

More specifically, the *EM* algorithm is an iterative method for locating posterior modes. Each iteration consists of two steps: The E-step (expectation step) and the M-step (maximization step). Formally, let θ^i denote the current guess to the mode of the observed posterior $p(\theta \mid Y)$; let $p(\theta \mid Y, Z)$ denote the augmented posterior, i.e. the posterior of the augmented data; and let $p(Z \mid \theta^i, Y)$ denote the conditional predictive distribution of the latent data Z, conditional on the current guess to the posterior mode. In the most general setting, the E-step consists of computing

$$Q(\theta, \theta^i) = \int_Z \log[p(\theta \mid Z, Y)] p(Z \mid \theta^i, Y) dZ,$$

i.e. the expectation of $\log p(\theta \mid Z, Y)$ with respect to $p(Z \mid \theta^i, Y)$. In the M-step the Q function is maximized with respect to θ to obtain θ^{i+1}. The algorithm is iterated until $\| \theta^{i+1} - \theta^i \|$ or $\mid Q(\theta^{i+1}, \theta^i) - Q(\theta^i, \theta^i) \mid$ is sufficiently small.

EXAMPLE. *Genetic Linkage Model (Rao, 1973)*

Suppose that 197 animals (Y) are distributed into four categories as follows:

$$Y = (y_1, y_2, y_3, y_4) = (125, 18, 20, 34)$$

with cell probabilities

$$\left(\frac{1}{2} + \frac{\theta}{4}, \frac{1}{4}(1 - \theta), \frac{1}{4}(1 - \theta), \frac{\theta}{4} \right).$$

Augment the observed data by splitting the first cell into two cells with probabilities $\frac{1}{2}$ and $\frac{\theta}{4}$. The augmented data are given by

$$(x_1, x_2, x_3, x_4, x_5)$$

such that

$$x_1 + x_2 = 125$$
$$x_3 = y_2$$
$$x_4 = y_3$$
$$x_5 = y_4.$$

Notice that the observed posterior (under a flat prior) is proportional to

$$(2 + \theta)^{y_1}(1 - \theta)^{y_2 + y_3}\theta^{y_4},$$

while the augmented posterior (under a flat prior) is proportional to

$$\theta^{x_2 + x_5}(1 - \theta)^{x_3 + x_4}.$$

By working with the augmented posterior, we realize a simplification in functional form.

For this genetic linkage model, note that

$$Q(\theta, \theta^i) = E[(x_2 + x_5)\log(\theta) + (x_3 + x_4)\log(1 - \theta) \mid \theta^i, Y],$$

where $p(Z \mid \theta^i, Y)$ is the binomial distribution with $n = 125$ and $p = \theta^i/(\theta^i + 2)$. In this case, $Q(\theta, \theta^i)$ simplifies to

$$[E(x_2 \mid \theta^i, Y) + x_5]\log(\theta) + (x_3 + x_4)\log(1 - \theta)$$

and is linear in the latent data.

For the M-step, note that

$$\frac{\partial Q(\theta, \theta^i)}{\partial \theta}\bigg|_{\hat{\theta}} = 0 \Rightarrow \frac{E(x_2 \mid \theta^i, Y) + x_5}{\hat{\theta}} - \frac{x_3 + x_4}{1 - \hat{\theta}} = 0$$

$$\Rightarrow \theta_{i+1} = \frac{E(x_2 \mid \theta^i, Y) + x_5}{E(x_2 \mid \theta^i, Y) + x_3 + x_4 + x_5}.$$

Starting at $\theta^\circ = 0.5$, the *EM* algorithm converges to $\theta^* = 0.6268$ (the observed posterior mode) after four iterations. While this example is quite simple, it is sufficiently rich to illustrate a number of algorithms to be presented in Chapters 5 and 6.

EXAMPLE. *Simple Linear Regression with Right-Censored Data.*
We consider the motorette data found in Schmee and Hahn (1979). Ten motorettes were tested at each of the four temperatures: 150°, 170°, 190° and 220°C. The time to failure in hours is given in Table 4.1.

A star indicates that a motorette was taken off the study without failing at the event time indicated. For these data, we will fit the model

$$t_i = \beta_0 + \beta_1 v_i + \sigma \varepsilon_i,$$

where $\varepsilon_i \sim N(0, 1)$; $v_i = 1000/(\text{temperature} + 273.2)$ and $t_i = \log_{10}$ (ith failure time).

Reorder the data so that the first m observations are uncensored (i.e. a failure is observed at t_i) and the remaining $n - m$ are censored (c_i denotes a censored event time). The log-augmented posterior (under a flat prior) is given (up to a constant by)

$$-n \log \sigma - \sum_{i=1}^{m} (t_i - \beta_0 - \beta_1 v_i)^2 / 2\sigma^2 - \sum_{i=m+1}^{n} (Z_i - \beta_0 - \beta_1 v_i)^2 / 2\sigma^2,$$

where Z_j is the (unobserved) failure time for case j. The conditional predictive distribution $p(Z_i \mid \beta_0, \beta_1, \sigma, c_i)$ is the conditional normal distribution, conditional on the fact that the unobserved failure time Z_j is greater than c_j. Hence, the E-step requires one to compute

$$-n \log \sigma - \frac{1}{2\sigma^2} \sum_{i=1}^{m} (t_i - \beta_0 - \beta_1 v_i)^2$$

$$- \frac{1}{2\sigma^2} \sum_{i=m+1}^{n} [E(Z_i^2 \mid \beta_0, \beta_1, \sigma, Z_i > c_i)$$

$$- 2(\beta_0 + \beta_1 v_i)E(Z_i \mid \beta_0, \beta_1, \sigma, Z_i > c_i) + (\beta_0 + \beta_1 v_i)^2].$$

TABLE 4.1.

150°	170°	190°	220°
8064*	1764	408	408
8064*	2772	408	408
8064*	3444	1344	504
8064*	3542	1344	504
8064*	3780	1440	504
8064*	4860	1680*	528*
8064*	5196	1680*	528*
8064*	5448*	1680*	528*
8064*	5448*	1680*	528*
8064*	5448*	1680*	528*

Source: Schmee and Hahn (1979).

Note that

$$E(Z_i^2 \mid \beta_0, \beta_1, \sigma, Z_i > c_i) = \mu_i^2 + \sigma^2 + \sigma(c_i + \mu_i) H\left(\frac{c_i - \mu_i}{\sigma}\right)$$

and

$$E(Z_i \mid \beta_0, \beta_1, \sigma, Z_i > c_i) = \mu_i + \sigma H\left(\frac{c_i - \mu_i}{\sigma}\right),$$

where $\mu_i = \beta_0 + \beta_1 v_i$, $H(x) = \phi(x)/\{1 - \Phi(x)\}$, and $\phi(x)$ and $\Phi(x)$ are the density and cdf of the standard normal distribution.

To see this last result, note that

$$E(Z_i \mid \beta_0, \beta_1, \sigma, Z_i > c_i)$$

$$= \beta_0 + \beta_1 v_i + \sigma E\left(\varepsilon_i \mid \varepsilon_i > \frac{c_i - \beta_0 - \beta_1 v_i}{\sigma}\right)$$

$$= \beta_0 + \beta_1 v_i + \sigma\left[\int_{\frac{c_i - \beta_0 - \beta_1 v_i}{\sigma}}^{\infty} w\phi(w)\,dw\right]\bigg/\left[1 - \Phi\left(\frac{c_i - \beta_0 - \beta_1 v_i}{\sigma}\right)\right]$$

$$= \beta_0 + \beta_1 v_i + \sigma\phi\left(\frac{c_i - \beta_0 - \beta_1 v_i}{\sigma}\right)\bigg/\left[1 - \Phi\left(\frac{c_i - \beta_0 - \beta_1 v_i}{\sigma}\right)\right]$$

$$= \beta_0 + \beta_1 v_i + \sigma H\left(\frac{c_i - \beta_0 - \beta_1 v_i}{\sigma}\right).$$

The *M*-step is then

$$\frac{\partial Q}{\partial \beta_0} = 0 \Rightarrow \sum_{i=1}^{m} (t_i - \beta_0 - \beta_1 v_i) + \sum_{i=m+1}^{n} \{E(Z_i) - \beta_0 - \beta_1 v_i\} = 0$$

$$\frac{\partial Q}{\partial \beta_1} = 0 \Rightarrow \sum_{i=1}^{m} v_i(t_i - \beta_0 - \beta_1 v_i) + \sum_{i=m+1}^{n} v_i\{E(Z_i) - \beta_0 - \beta_1 v_i\} = 0$$

$$\frac{\partial Q}{\partial \sigma^2} = 0 \Rightarrow \frac{\sum_{i=1}^{m} (t_i - \beta_0 - \beta_1 v_i)^2}{\sigma^4}$$

$$+ \frac{\sum_{i=m+1}^{n} E(Z_i^2) - 2(\beta_0 + \beta_1 v_i)E(Z_i) + (\beta_0 + \beta_1 v_i)^2}{\sigma^4} - \frac{n}{\sigma^2} = 0.$$

To obtain β^{i+1}, replace c_j by $E(Z_j \mid \beta_0^i, \beta_1^i, \sigma^i, Z_j > c_j)$ and apply least squares. To obtain σ_{i+1}^2, compute

$$\sigma_{i+1}^2 = \frac{\sum_{j=1}^{m} (t_j - \mu_j^i)^2}{n} + \sigma_i^2 \frac{\sum_{j=m+1}^{n}\left[1 + \left(\frac{c_j - u_j^i}{\sigma_i}\right) H\left(\frac{c_j - \mu_j^i}{\sigma_i}\right)\right]}{n},$$

where $\mu_j^i = \beta_0^i + \beta_1^i v_j$.

Applying the *EM* algorithm to the data of Schmee and Hahn (1979), we arrive at

$$\hat{\beta}_0 = -6.019, \quad \hat{\beta}_1 = 4.311 \quad \text{and} \quad \hat{\sigma} = 0.2592$$

after 16 iterations.

EXAMPLE. *Item Response Model*

In Section 2.3.2, we introduced the item response model and presented the Newton–Raphson solution for locating the mode of the ζ marginal of the posterior distribution.

Treating the unobserved ability parameters $\theta = (\theta_1, \ldots, \theta_n)$ as latent data Z and $\zeta = (\zeta_1, \ldots, \zeta_k)$ as the parameter, the Q function is given as

$$Q(\zeta, \zeta^*) = \int \log p(\zeta \mid \theta, Y) p(\theta \mid Y, \zeta^*) d\theta,$$

which in turn is equal to

$$\sum_{i=1}^{n} \sum_{j=1}^{k} \int \log \left[p_{ij}^{y_{ij}} (1 - p_{ij})^{1-y_{ij}} \right] p(\theta_i \mid Y, \zeta^*) d\theta_i$$

$$+ \sum_{i=1}^{n} \int \log[\phi(\theta_i)] p(\theta_i \mid Y, \zeta^*) d\theta_i, \qquad (*),$$

where $p(\theta_i \mid Y, \zeta^*)$ is proportional to $\phi(\theta_i) \prod_{j=1}^{k} p_{ij}^{y_{ij}} (t - p_{ij})^{1-y_{ij}}$ and $\phi(.)$ is the standard normal density. Note that the second term in (*) does not depend on ζ and, therefore, need not be computed. Thus, it can be shown that

$$\frac{\partial Q(\zeta, \zeta^*)}{\partial \zeta} = 0$$

implies that

$$\sum_{i=1}^{n} y_{ij} = \sum_{i=1}^{n} \int \frac{\exp(\alpha_j \theta_i - \beta_j) p(\theta_i \mid Y, \zeta^*)}{1 + \exp(\alpha_j \theta_i - \beta_j)} d\theta_i$$

$$\sum_{i=1}^{n} y_{ij} \int \theta_i p(\theta_i \mid Y, \zeta^*) d\theta_i = \sum_{i=1}^{n} \int \frac{\theta_i \exp(\alpha_j \theta_i - \beta_j) p(\theta_i \mid Y, \zeta^*)}{1 + \exp(\alpha_j \theta_i - \beta_j)} d\theta_i,$$

for $j = 1, \ldots, k$. Thus, rather than have $2k$ unknowns in $2k$ equations (as is the case for the Newton–Raphson solution), the *EM* solution consists of k independent sets of 2 unknowns in 2 equations. This follows since the *EM* solution uses $p(\theta_i \mid Y, \zeta^*)$ rather than $p(\theta_i \mid Y, \zeta)$, which is a function of the entire vector of parameters $(\zeta_1, \ldots, \zeta_k)$. Note, however, that like the Newton–Raphson solution, the *EM* solution requires numerical integration with respect to the θ_i's.

4.2 Theory

To begin, notice that

$$\log[p(\theta \mid Y)] = \log[p(\theta \mid Z, Y)] - \log[p(Z \mid \theta, Y)] + \log[p(Z \mid Y)].$$

Now integrate both sides of this equation with respect to the conditional predictive distribution $p(Z \mid Y, \phi)$ to obtain

$$\log[p(\theta \mid Y)] = \int_Z \log[p(\theta \mid Y, Z)]p(Z \mid \phi, Y)dZ$$

$$- \int_Z \log[p(Z \mid \theta, Y)]p(Z \mid \phi, Y)dZ$$

$$+ \int_Z \log[p(Z \mid Y)]p(Z \mid \phi, Y)dZ.$$

Dempster, Laird and Rubin (1977) define the Q function by

$$Q(\theta, \theta^*) = \int_Z \log[p(\theta \mid Z, Y)]p(Z \mid \theta^*, Y)dZ$$

and the H function by

$$H(\theta, \theta^*) = \int_Z \log[p(Z \mid \theta, Y)]p(Z \mid \theta^*, Y)dZ.$$

Let $K(\theta, \theta^*) = \int_Z \log[p(Z \mid Y)]p(Z \mid \theta^*, Y)dZ$. Note that

$$\log[p(\theta^{i+1} \mid Y)] - \log[p(\theta^i \mid Y)] = Q(\theta^{i+1}, \theta^i) - Q(\theta^i, \theta^i)$$

$$- [H(\theta^{i+1}, \theta^i) - H(\theta^i, \theta^i)] + \{K(\theta^{i+1}, \theta^i) - K(\theta^i, \theta^i)\}.$$

Dempster, Laird and Rubin (1977) define an *EM* algorithm to select θ^{i+1} so that $Q(\theta, \theta^i)$ is maximized as a function of θ. These authors define a *GEM* (generalized *EM*) algorithm to select θ^{i+1} so that $Q(\theta^{i+1}, \theta^i) > Q(\theta^i, \theta^i)$. From formula (le.6.6) of Rao (1973), it follows that

$$H(\theta^{i+1}, \theta^i) - H(\theta^i, \theta^i) \le 0.$$

Moreover, $K(\theta^{i+1}, \theta^i) - K(\theta^i, \theta^i) = 0$. Hence, we have shown:

Theorem 4.2.1. Every *EM* or *GEM* algorithm increases the posterior $p(\theta \mid Y)$ at each iteration, i.e. $p(\theta^{i+1} \mid Y) \ge p(\theta^i \mid Y)$, equality holding iff $Q(\theta^{i+1}, \theta^i) = Q(\theta^i, \theta^i)$.

In addition, the following results are available. See Wu (1983) and Little and Rubin (1987) for further details.

Theorem 4.2.2. Suppose a sequence of *EM* iterates θ^i satisfy

1. $\frac{\partial Q(\theta, \theta^i)}{\partial \theta} \big|_{\theta=\theta^{i+1}} = 0.$
2. θ^i converge to some value θ^* and let $p(Z \mid Y, \theta)$ be "sufficiently" smooth.

Then it follows that

$$\frac{\partial \log \ p(\theta \mid Y)}{\partial \theta}\Bigg|_{\theta=\theta*} = 0.$$

In other words, if the iterates θ^i converge, they converge to a stationary point of $p(\theta \mid Y)$. This implies that when there are multiple stationary points (local maxima or saddle points), the algorithm may not converge to the global maximum.

Dempster, Laird and Rubin (1977) show that the *EM* algorithm converges at a linear rate, with the rate depending on the proportion of information about θ in $p(\theta \mid Y)$ that is observed. This implies that the convergence can be quite slow if a large portion of the data are missing. A more rigorous treatment of this statement is given in Section 4.4.5.

4.3 *EM* in the Exponential Family

Suppose that the augmented data $X = (Y, Z)$ are distributed as

$$f(X \mid \theta) = b(X) \exp[\theta^T s(X)]/\alpha(\theta),$$

i.e. the regular exponential family, where θ is a $d \times 1$ vector, and $s(X)$ is a $1 \times d$ vector of sufficient statistics. In this case, the Q function is given as

$$Q(\theta, \theta^i) = \int \log \ p(\theta \mid Y, Z)p(Z \mid Y, \theta^i)dZ$$

$$= \int_Z \{\log[b(X)] + \theta^T s(X) - \log[a(\theta)]\}p(Z \mid \theta^i, Y)dZ$$

$$= \int_Z \log[b(X)]p(Z \mid Y, \theta^i)dZ + \theta^T \int_Z s(X)p(Z \mid Y, \theta^i)dZ - \log[a(\theta)].$$

Now notice that

$$\int_Z \log[b(X)]p(Z \mid \theta^i, Y)dZ$$

does not depend on θ, so there is no need to compute it in the E-step. Thus, the E-step consists of computing

$$E[s(X) \mid Y, \theta^i] = \int_Z s(X)p(Z \mid Y, \theta^i)dZ = s^i.$$

The M-step consists of maximizing $Q(\theta, \theta^i)$ or

$$- \log[a(\theta)] + \theta^T s^i.$$

In this way,

$$\frac{\partial Q(\theta, \theta^i)}{\partial \theta} = 0 \Rightarrow \frac{-\partial \ \log[a(\theta)] - \theta^T s^i}{\partial \theta} = 0 \Rightarrow \frac{\partial \ \log \ a(\theta)}{\partial \theta} = s^i.$$

Now recall that

$$a(\theta) = \int_X b(X) \exp\{\theta^T s(X)\}dX.$$

So

$$\frac{\partial \log a(\theta)}{\partial \theta} = \left\{ \int_X b(X) \frac{\partial \exp[\theta^T s(X)]}{\partial \theta} dX \right\} \Big/ a(\theta)$$

$$= \left\{ \int_X s(X) b(X) \exp[\theta^T s(X)] dX \right\} \Big/ a(\theta)$$

$$= E[s(X) \mid \theta].$$

Thus, in the regular exponential family, maximizing $Q(\theta, \theta')$ is equivalent to solving

$$E[s(X) \mid \theta] = s^i$$

for θ.

EXAMPLE. *Variance Components*
Consider the 1-way model

$$y_{ij} = \beta + \alpha_i + \varepsilon_{ij},$$

where y_{ij} is the jth observation from the ith class, β is the grand mean, α_i is the effect of the ith group and ε_{ij} is the error, where $i = 1, \ldots, a, j = 1, \ldots, n_i$ and $\sum_{i=1}^{a} n_i = N$. In matrix notation, we have

$$y = 1\beta + Z\alpha + \varepsilon,$$

where 1 is an $N \times 1$ vector of 1's and

$$Z = \begin{bmatrix} 1 & 0 & . & . & . & 0 \\ . & . & . & . & . & . \\ . & . & . & . & . & . \\ 1 & 0 & . & . & . & 0 \\ 0 & 1 & . & . & . & 0 \\ . & . & . & . & . & . \\ . & . & . & . & . & . \\ 0 & 1 & . & . & . & 0 \\ . & . & . & . & . & . \\ . & . & . & . & . & . \\ 0 & 0 & . & . & . & 1 \\ . & . & . & . & . & . \\ . & . & . & . & . & . \\ 0 & 0 & . & . & . & 1 \end{bmatrix},$$

i.e. an $N \times a$ matrix, with n_k 1's in the kth column. Assuming that the errors follow the $N(0, \sigma_\varepsilon^2 I_N)$ distribution, where I_N is the $N \times N$ identity matrix, and α follows the $N(0, \sigma_\alpha^2 I_a)$ distribution, where I_a is the $a \times a$ identity matrix, with ε and α independent, we have

$$V = \mathrm{Var}(y) = ZZ^T \sigma_\alpha^2 + I_N \sigma_\varepsilon^2,$$

and

$$\mathrm{cov}(y, \alpha^T) = \mathrm{cov}(1\beta + Z\alpha + \varepsilon, \alpha^T) = Z\mathrm{cov}(\alpha, \alpha^T) = Z\sigma_\alpha^2.$$

Treating α as the latent data, the joint distribution of y and α (i.e. the augmented likelihood) is $N(\mu, \Sigma)$, where

$$\mu = \begin{bmatrix} 1\beta \\ 0 \\ \cdot \\ \cdot \\ \cdot \\ 0 \end{bmatrix}$$

and

$$\Sigma = \begin{bmatrix} V & \sigma_\alpha^2 Z \\ \sigma_\alpha^2 Z & \sigma_\alpha^2 I_a \end{bmatrix}.$$

The sufficient statistics and their unconditional expectations are

1. $\alpha^T \alpha / a$; σ_α^2
2. $\varepsilon^T \varepsilon / N$; σ_ε^2
3. $1^T (y - Z\alpha)/N$; β.

Searle, Casella and McCulloch (1992) show that the conditional expectations of $\alpha^T \alpha$, $\varepsilon^T \varepsilon$ and $1^T(y - Z\alpha)$ are, respectively,

$$\tau^2 \sum_{i=1}^{a} \frac{(\bar{y}_{i.} - \beta)^2}{(\tau + 1/n_i)^2} + a\sigma_\alpha^2 - \sigma_\alpha^2 \tau \sum_{i=1}^{a} \frac{1}{\tau + 1/n_i}$$

$$\sum_{i=1}^{a} \sum_{j=1}^{n_i} (y_{ij} - \beta)^2 - \tau \sum_{i=1}^{a} \frac{(\bar{y}_{i.} - \beta)^2}{(\tau + 1/n_i)^2}(2 + n_i\tau) + \sigma_\varepsilon^2 \sum_{i=1}^{a} \frac{\tau}{\tau + 1/n_i}$$

$$\sum_{i=1}^{a} \frac{\bar{y}_{i.} + n_i \tau \beta}{\tau + 1/n_i},$$

where

$$\tau = \frac{\sigma_\alpha^2}{\sigma_\varepsilon^2}$$

and $\bar{y}_{i.}$ is the average response in group i.

Variance component models for complex pedigrees are discussed in Thompson and Shaw (1990) and Guo and Thompson (1991; 1994). Searle, Casella and McCulloch (1992) discuss the related REML technique.

4.4 Standard Errors in the Context of *EM*

The output of the *EM* algorithm is the mode of the posterior distribution $p(\theta \mid Y)$. To define the normal approximation to $p(\theta \mid Y)$, one must compute the Hessian (matrix) of $\log p(\theta \mid Y)$. Several approaches are available, which are now discussed.

4.4.1. Direct Computation/Numerical Differentiation

Having arrived at the observed posterior mode, θ^*, one can potentially evaluate

$$-\frac{\partial^2 \log p(\theta \mid Y)}{\partial \theta^2}\bigg|_{\theta^*}.$$

In practice, however, this may be tedious to code or difficult to evaluate for a given data set.

An alternative approach is to numerically differentiate $\partial \log p(\theta \mid Y)/\partial\theta$ to obtain the Hessian. Meilijson (1989) suggests that one can perturb $\theta^* = (\theta_1^*, \ldots, \theta_d^*)$ by adding (a small amount) $\varepsilon > 0$ to one coordinate and evaluate $\partial \log p(\theta \mid Y)/\partial\theta$ at the perturbed parameter $\tilde{\theta}$. The ith row of the Hessian is approximately

$$\frac{1}{\varepsilon}\left[\frac{\partial \log p(\theta \mid Y)}{\partial\theta}\bigg|_{\tilde{\theta}} - \frac{\partial \log p(\theta \mid Y)}{\partial\theta}\bigg|_{\theta^*}\right].$$

The entire Hessian requires d such operations. Care must be taken in the selection of ε to ensure that the derivatives are not distorted by the round-off errors.

4.4.2. Missing Information Principle

Note that

$$\log p(\theta \mid Y) = \log p(\theta \mid Y, Z) - \log p(Z \mid Y, \theta) + C.$$

This implies that

$$\frac{-\partial^2 \log p(\theta \mid Y)}{\partial \theta^2} = -\frac{\partial^2 \log p(\theta \mid Y, Z)}{\partial \theta^2} + \frac{\partial^2 \log p(Z \mid Y, \theta)}{\partial \theta^2}.$$

Integrating both sides of this equation with respect to $p(Z \mid Y, \theta)$, we obtain

$$-\frac{\partial^2 \log p(\theta \mid Y)}{\partial \theta^2} = -\int_Z \frac{\partial^2 \log p(\theta \mid Y, Z)}{\partial \theta^2} p(Z \mid \theta, Y)dZ$$

$$+ \int_Z \frac{\partial^2 \log p(Z \mid Y, \theta)}{\partial \theta^2} p(Z \mid Y, \theta)dZ$$

$$= \frac{-\partial^2 Q(\theta, \phi)}{\partial \theta^2}\bigg|_{\phi=\theta} - \frac{-\partial^2 H(\theta, \phi)}{\partial \theta^2}\bigg|_{\phi=\theta}, \tag{4.4.1}$$

To prove Louis' result, note that

$$\frac{\partial^2 \log p(\theta \mid Y)}{\partial \theta^2} = \frac{\partial \int_Z (\partial \log p(\theta \mid Y, Z)/\partial \theta) p(Z \mid Y, \theta) dZ}{\partial \theta}$$

$$= \int_Z \frac{\partial^2 \log p(\theta \mid Y, Z)}{\partial \theta^2} p(Z \mid Y, \theta) dZ$$

$$+ \int_Z \frac{\partial \log p(\theta \mid Y, Z)}{\partial \theta} \frac{\partial p(Z \mid Y, \theta)}{\partial \theta} dZ$$

$$= (^*) + (\Delta)$$

by the chain rule. Now

$$\Delta = \int_Z \frac{\partial \log p(\theta \mid Y, Z)}{\partial \theta} \frac{[\partial p(Z \mid Y, \theta)/\partial \theta]}{p(Z \mid Y, \theta)} p(Z \mid Y, \theta) dZ$$

$$= \int_Z \frac{\partial \log p(\theta \mid Y, Z)}{\partial \theta} \frac{\partial \log p(Z \mid Y, \theta)}{\partial \theta} p(Z \mid Y, \theta) dZ.$$

Since

$$\frac{\partial \log p(Z \mid Y, \theta)}{\partial \theta} = \frac{\partial \log p(\theta \mid Y, Z)}{\partial \theta} - \frac{\partial \log p(\theta \mid Y)}{\partial \theta},$$

it follows that

$$\Delta = \int_Z \frac{\partial \log p(\theta \mid Y, Z)}{\partial \theta} \left[\frac{\partial \log p(\theta \mid Y, Z)}{\partial \theta} - \frac{\partial \log p(\theta \mid Y)}{\partial \theta} \right] \times$$

$$p(Z \mid Y, \theta) dZ$$

$$= \int_Z \left(\frac{\partial \log p(\theta \mid Y, Z)}{\partial \theta} \right)^2 p(Z \mid Y, \theta) dZ$$

$$- \left[\int_Z \frac{\partial \log p(\theta \mid Y, Z)}{\partial \theta} p(Z \mid Y, \theta) dZ \right]^2,$$

since

$$\int_Z \frac{\partial \log p(\theta \mid Y, Z)}{\partial \theta} p(Z \mid Y, \theta) dZ = \frac{\partial \log p(\theta \mid Y)}{\partial \theta}.$$

Note that (4.4.1) and (4.4.2) imply that $\Delta = -\partial^2 H(\theta, \phi)/\partial \theta^2 \mid_{\phi=\theta}$.

Finally, note that at the posterior mode $\hat{\theta}$, $\partial \log p(\theta \mid Y)/\partial \theta \mid_{\hat{\theta}} = 0$. Hence, at $\hat{\theta}$

$$\Delta = \int_Z \left(\frac{\partial \log p(\theta \mid Y, Z)}{\partial \theta} \right)^2 p(Z \mid Y, \hat{\theta}) dZ.$$

EXAMPLE. *Genetic Linkage (Continued)*
For this model,

$$\frac{\partial \log p(\theta \mid Y, Z)}{\partial \theta} = \frac{x_2 + x_5}{\theta} - \frac{x_3 + x_4}{1 - \theta}$$

where Q and H are defined in Section 4.2. Referring to $-\partial^2 Q/\partial\theta^2$ as the complete information and to $-\partial^2 H/\partial\theta^2$ as the missing information, we have the Missing Information Principle (Louis, 1982):

Observed Information = Complete Information − Missing Information.

Orchard and Woodbury (1972) present an analogous expression for expected information by taking expectations over Y.

4.4.3. Louis' Method

A basic result due to Louis (1982) is that $-\partial^2 H/\partial\theta^2 = \mathrm{var}[\partial \log p(\theta \mid Y, Z)/\partial\theta]$. Thus, it follows that

$$
\frac{-\partial^2 \log p(\theta \mid Y)}{\partial\theta^2} = -\int_Z \frac{\partial^2 \log p(\theta \mid Y, Z)}{\partial\theta^2} p(Z \mid Y, \theta) dZ
$$

$$
- \mathrm{var}\left\{ \frac{\partial \log p(\theta \mid Y, Z)}{\partial\theta} \right\}, \tag{4.4.2}
$$

where the variance is with respect to $p(Z \mid Y, \theta)$. We now prove this result. Those readers not interested in the proof may skip to the example.

To prove Louis' result, we first need to show

$$
\frac{\partial \log p(\theta \mid Y)}{\partial\theta} = \int_Z \frac{\partial \log p(\theta \mid Y, Z)}{\partial\theta} p(Z \mid \theta, Y) dZ, \tag{4.4.3}
$$

i.e. the observed score is equal to the expected score of the log-augmented posterior. To see this, note that

$$
\frac{\partial \log p(\theta \mid Y)}{\partial\theta} = \frac{\partial \log p(\theta \mid Y, Z)}{\partial\theta} - \frac{\partial \log p(Z \mid Y, \theta)}{\partial\theta},
$$

which implies that

$$
\frac{\partial \log p(\theta \mid Y)}{\partial\theta} = \int_Z \frac{\partial \log p(\theta \mid Y, Z)}{\partial\theta} p(Z \mid Y, \theta) dZ
$$

$$
- \int_Z \frac{\partial \log p(Z \mid Y, \theta)}{\partial\theta} p(Z \mid Y, \theta) dZ.
$$

Regarding the last term,

$$
\int_Z \frac{\partial \log p(Z \mid Y, \theta)}{\partial\theta} p(Z \mid Y, \theta) dZ = \int_Z \frac{\partial p(Z \mid Y, \theta)}{\partial\theta} dZ.
$$

Interchanging derivative and integral,

$$
\int_Z \frac{\partial p(Z \mid Y, \theta)}{\partial\theta} dZ = \frac{\partial \int_Z p(Z \mid Y, \theta) dZ}{\partial\theta} = 0,
$$

since $p(Z \mid Y, \theta)$ integrates to unity over the space Z. The result (4.4.3) follows.

while

$$\left.\frac{-\partial^2\ Q(\theta,\hat{\theta})}{\partial\theta^2}\right|_{\hat{\theta}} = \frac{E(x_2\mid\hat{\theta},Y)+x_5}{\hat{\theta}^2} + \frac{x_3+x_4}{(1-\hat{\theta})^2}.$$

For the data set listed in the sample of Section 4.1

$$\left.-\frac{\partial^2 Q(\theta,\hat{\theta})}{\partial\theta^2}\right|_{\hat{\theta}} = \frac{(29.83+34)}{0.6268^2} + \frac{38}{(1-0.6268)^2} = 435.3.$$

Now

$$\mathrm{var}\left(\left.\frac{\partial\ \log\ p(\theta\mid Y,Z)}{\partial\theta}\right|_{\hat{\theta}}\right) = \frac{\mathrm{var}(x_2\mid\hat{\theta})}{\hat{\theta}^2} = 125\left(\frac{\hat{\theta}}{2+\hat{\theta}}\right)\left(\frac{2}{2+\hat{\theta}}\right)\Big/\hat{\theta}^2$$

$$= 22.71/0.6268^2 = 57.8.$$

Hence, it follows that

$$\left.\frac{-\partial^2\ \log\ p(\theta\mid Y)}{\partial\theta^2}\right|_{\hat{\theta}} = 435.3 - 57.8 = 377.5$$

and the standard error of $\hat{\theta}$ is equal to $\sqrt{(1/377.5)} = 0.05$.

4.4.4. Simulation

In some situations, it may be difficult to analytically compute

$$\int_Z \frac{\partial^2\ \log\ p(\theta\mid Y,Z)}{\partial\theta^2}\ p(Z\mid\hat{\theta},Y)dZ.$$

If one can sample from $p(Z\mid\theta,Y)$, this integral may be approximated by the sum

$$\frac{1}{m}\sum_{j=1}^m \frac{\partial^2\ \log\ p(\theta\mid Y,z_j)}{\partial\theta^2},$$

given $z_1, z_2, \ldots, z_m \overset{iid}{\sim} p(Z\mid\hat{\theta},Y)$, Rubin (1987b) refers to the z_i's as multiple imputations.

Similarly, one can approximate

$$\mathrm{var}\left(\left.\frac{\partial\ \log\ p(\theta\mid Y,Z)}{\partial\theta}\right|_{\hat{\theta}}\right)$$

via

$$\frac{1}{m}\sum_{j=1}^m \left(\frac{\partial\ \log\ p(\theta\mid Y,z_j)}{\partial\theta}\right)^2 - \left[\frac{1}{m}\sum_{j=1}^m \frac{\partial\ \log\ p(\theta\mid Y,z_j)}{\partial\theta}\right]^2,$$

where

$$z_1, \ldots, z_m \sim p(Z\mid\hat{\theta},Y).$$

EXAMPLE. *Genetic Linkage (Continued)*
In the genetic linkage example, $p(Z \mid \hat{\theta}, Y)$ is the binomial distribution with parameters $n = 125$ and $p = \hat{\theta}/(2 + \hat{\theta})$. Having converged to the posterior mode $\hat{\theta} = 0.6268$, 10,000 imputations were drawn from the binomial distribution with parameters $n = 125$ and $p = \hat{\theta}/(\hat{\theta} + 2)$. Computations were performed using the RANDOM command of MINITAB. The estimated variance of the augmented score

$$\frac{1}{10,000} \sum_{j=1}^{10,000} \left(\frac{\partial \ \log \ p(\theta \mid Y, z_j)}{\partial \theta} \right)^2$$

was found to be equal to 57.4. The exact value is 57.8.

4.4.5. Using EM Iterates

In this approach, we show how to adjust the Hessian of the Q function using the *EM* iterates θ^i to obtain the variance–covariance matrix. A one-dimensional version of this approach is given in Smith (1977). The multivariate version is given in Meng and Rubin (1991).

The *EM* algorithm defines a mapping $M : \theta \rightarrow \theta$ from the parameter space θ onto itself such that

$$\theta^{i+1} = M(\theta^i).$$

If the *EM* iterates converge to θ^*, then

$$\theta^* = M(\theta^*).$$

Dempster, Laird and Rubin (1977) show that

$$\left\{ \frac{\partial \ M(\theta)}{\partial \theta} \bigg|_{\theta*} \right\} \left\{ \frac{\partial^2 \ Q(\theta, \theta^*)}{\partial \theta^2} \bigg|_{\theta*} \right\} = \frac{\partial^2 \ H(\theta, \theta^*)}{\partial \theta^2} \bigg|_{\theta*}. \tag{4.4.4}$$

These authors use this result to argue that in a neighborhood of θ^*, the rate of convergence of the *EM* algorithm is given by:

$$\left(\frac{\partial^2 \ H}{\partial \theta^2} \right) \left(\frac{\partial^2 Q}{\partial \theta^2} \right)^{-1},$$

i.e. the ratio of the missing information to the complete information.
From (4.4.4) and the decomposition

$$\frac{-\partial^2 \ \log \ p(\theta \mid Y)}{\partial \theta^2} \bigg|_{\theta*} = \frac{-\partial^2 \ Q(\theta, \theta^*)}{\partial \theta^2} \bigg|_{\theta*} - \frac{-\partial^2 \ H(\theta, \theta^*)}{\partial \theta^2} \bigg|_{\theta*}$$

given in Section 4.4.2, it follows that

$$\frac{-\partial^2 \ \log \ p(\theta \mid Y)}{\partial \theta^2} \bigg|_{\theta*} = \left[I - \frac{\partial \ M(\theta)}{\partial \theta} \bigg|_{\theta*} \right] \left[\frac{-\partial^2 \ Q(\theta, \theta^*)}{\partial \theta^2} \bigg|_{\theta*} \right],$$

where I is the $d \times d$ identity matrix. It is easy to check that

$$
\left[\frac{-\partial^2 \log p(\theta \mid Y)}{\partial\theta^2} \bigg|_{\theta*} \right]^{-1} = \left[\frac{-\partial^2 Q(\theta,\theta*)}{\partial\theta^2} \bigg|_{\theta*} \right]^{-1} + \left[\frac{\partial^2 Q(\theta,\theta*)}{\partial\theta^2} \bigg|_{\theta*} \right]^{-1}
$$

$$
\times \left[I - \frac{\partial M(\theta)}{\partial\theta} \bigg|_{\theta*} \right]^{-1} \left[\frac{\partial M(\theta)}{\partial\theta} \bigg|_{\theta*} \right].
$$

Meng and Rubin (1991) note that

$$
r_{ij} = \frac{\partial M_j(\theta)}{\partial\theta_i} \bigg|_{\theta*} = \lim_{\theta_i \to \theta_i^*} \frac{M_j(\theta_1^*, \ldots, \theta_i, \ldots, \theta_d^*) - M_j(\theta*)}{\theta_i - \theta_i^*}
$$

$$
= \lim_{t \to \infty} \frac{M_j(\theta_1^*, \ldots, \theta_i^t, \ldots, \theta_d^*) - (\theta_j^*)}{\theta_i^t - \theta_i^*} = \lim_{t \to \infty} r_{ij}^t.
$$

In other words, these authors use the *EM* iterates to differentiate numerically $M(\theta)$. In this way, to compute r_{ij}^t, having obtained $\theta*$ and θ^t, Meng and Rubin (1991) suggest:

1. Fix $i = 1$ and form $\theta^t(i) = (\theta_1^*, \ldots, \theta_i^t, \ldots, \theta_d^*)$. Evaluate $\tilde{\theta}^{t+1}(i) = M[\theta^t(i)]$.
2. Form

$$
r_{ij}^t = \frac{\tilde{\theta}_j^{t+1}(i) - \theta_j^*}{\theta_i^t - \theta_i^*}
$$

   for $j = 1, \ldots, d$.
3. Repeat steps 1 and 2 for $i = 2, \ldots, d$.

To implement these three steps, d evaluations of the mapping M are required. Meng and Rubin (1991) obtain r_{ij} when the sequence $r_{ij}^{t^*}, r_{ij}^{t^*+1}, \ldots$ is stable for some t^*. As these authors illustrate, different values of t^* may be used for different r_{ij} elements, but care must be taken in coding to avoid problems with round-off error in computing the finite difference in step 3. In addition, Meng and Rubin (1991) point out that the present estimate of $\partial^2 \log p(\theta \mid Y)/\partial\theta^2 \mid_{\hat{\theta}}$ may not be symmetric. A possible solution to this problem would be to replace the estimated variance–covariance matrix $\hat{V}$ with $\frac{1}{2}(\hat{V} + \hat{V}^T)$, though this idea has not been investigated.

Meng and Rubin argue that the present approach will be of use if calculation of the variance of the augmented score (Approach 3) is tedious or intractable. In such a case, it would be of interest to compare the present approach with the simulation technique presented in Section 4.4.4. Clearly, the simulation approach moves out of the context of the classic *EM* as it require access to a random number generator.

To prove (4.4.4), first note that

$$
\frac{\partial Q(\theta, \theta_1)}{\partial \theta}\bigg|_{\theta_2} = \frac{\partial Q(\theta, \theta^*)}{\partial \theta}\bigg|_{\theta^*} + (\theta_2 - \theta^*)\frac{\partial^2 Q(\theta, \theta^*)}{\partial \theta^2}\bigg|_{\theta^*}
$$

$$
+ (\theta_1 - \theta^*)\frac{\partial Q(\theta, \phi)}{\partial \theta \partial \phi}\bigg|_{\substack{\theta=\theta^* \\ \phi=\theta^*}} + \cdots .
$$

Substituting $\theta_1 = \theta'$ and $\theta_2 = \theta'^{+1}$, we have

$$
0 = (\theta'^{+1} - \theta^*)\frac{\partial^2 Q(\theta, \theta^*)}{\partial \theta^2}\bigg|_{\theta^*} + (\theta' - \theta^*)\frac{\partial Q(\theta, \phi)}{\partial \theta \partial \phi}\bigg|_{\substack{\theta=\theta^* \\ \phi=\theta^*}} + \cdots .
$$

Since $\theta'^{+1} = M(\theta')$ and $\theta^* = M(\theta^*)$, we have in the limit

$$
0 = \left[\frac{\partial M(\theta)}{\partial \theta}\bigg|_{\theta^*}\right]\left[\frac{\partial^2 Q(\theta, \theta^*)}{\partial \theta^2}\bigg|_{\theta^*}\right] + \frac{\partial Q(\theta, \phi)}{\partial \theta \partial \phi}\bigg|_{\substack{\theta=\theta^* \\ \phi=\theta^*}} .
$$

However, as noted in Section 4.4.3,

$$
\frac{\partial Q(\theta, \phi)}{\partial \theta \partial \phi}\bigg|_{\substack{\theta=\theta^* \\ \phi=\theta^*}} = \frac{-\partial^2 H(\theta, \theta^*)}{\partial \theta^2}\bigg|_{\theta^*} .
$$

Hence,

$$
\left[\frac{\partial M(\theta)}{\partial \theta}\bigg|_{\theta^*}\right]\left[\frac{\partial^2 Q(\theta, \theta^*)}{\partial \theta^2}\bigg|_{\theta^*}\right] = \frac{\partial^2 H(\theta, \theta^*)}{\partial \theta^2}\bigg|_{\theta^*} .
$$

4.5 Monte Carlo Implementation of the *E*-Step

Given the current guess to the posterior mode, θ^i, the *E*-step requires the computation

$$
Q(\theta, \theta^i) = \int_Z \log p(\theta \mid Y, Z)p(Z \mid \theta^i, Y)dZ.
$$

To facilitate the *E*-step, one may supply the method of Monte Carlo to calculate the Q function. In particular, the Monte Carlo *E*-step (Wei and Tanner, 1990a) is given as

a. Draw $z_1, \ldots, z_m \overset{iid}{\sim} p(Z \mid Y, \theta^i)$.
b. Let $\hat{Q}_{i+1}(\theta, \theta^i) = \frac{1}{m}\sum_{j=1}^{m} \log p(\theta \mid z_j, Y)$.

In the *M*-step, $\hat{Q}$ is maximized to obtain θ^{i+1}.

Two important considerations in implementing Monte Carlo *EM* are the specification of m and the monitoring of convergence. Regarding the specification of m, it is inefficient to start with a large value of m when θ^i is far from the mode.

Rather, one may increase m as the current approximation moves closer to the true value (θ^*).

One may monitor the convergence of the algorithm by plotting θ^i vs. iteration number i. After a certain number of iterations, the plot will reveal random fluctuation about the line $\theta = \theta^*$. At such a point, one may terminate the algorithm or continue with a large value of m to decrease the system variability.

Note that within the context of Monte Carlo EM, the observed information matrix of the observed data at the posterior mode θ^* can be estimated via

$$\frac{1}{m} \sum_{j=1}^{m} \frac{\partial^2 \log\ p(\theta \mid Y, z_i)}{\partial\theta^2} \Bigg|_{\theta*} + \frac{1}{m} \sum_{j=1}^{m} \left(\frac{\partial\ \log\ p(\theta \mid Y, z_j)}{\partial\theta} \Bigg|_{\theta*} \right)^2 .$$

As will be seen in Section 5.5, a slight modification of Monte Carlo EM is available—the poor man's data augmentation algorithm—which results in an approximation to the entire posterior distribution, rather than just the maximizer. Chan and Ledholter (1995) present regularity conditions under which a $MCEM$ algorithm will, with high probability, converge to a maximizer of the likelihood or posterior.

EXAMPLE. *Genetic Linkage (Continued)*
Given the current guess to the posterior mode, θ^i, the conditional predictive distribution $p(Z \mid \theta^i, Y)$ is the binomial distribution with $n = 125$ and $p = \theta^i/(2 + \theta^i)$. As noted in the example of Section 4.1, the log-augmented posterior log $p(\theta \mid Y, Z)$, under a flat prior, is

$$(Z + x_5) \log(\theta) + (x_3 + x_4) \log(1 - \theta).$$

Having sampled $z_1, \ldots, z_m \overset{iid}{\sim} Bi(125; \theta^i/(2 + \theta^i))$, the Q function is calculated as

$$(ave + x_5) \log(\theta) + (x_3 + x_4) \log(1 - \theta),$$

where $ave = \Sigma_{j=1}^{m}\ z_j/m$. It follows that

$$\theta^{i+1} = \frac{ave + x_5}{ave + x_5 + x_3 + x_4} .$$

Table 4.2 presents the history of an implementation of the Monte Carlo EM algorithm for these data. The algorithm was initiated at $\theta^0 = 0.4$ and m was equal to 10 (1000) for iterations 1–8 (9–12).

As can be seen from Table 4.2, the process with $m = 10$ stabilized by the eighth iteration. From iterations 9–12, the mode is found to be equal to 0.627. The mode to four places is 0.6268.

TABLE 4.2.   History of *MCEM*

Iteration	Theta
1	0.5833
2	0.6222
3	0.6192
4	0.6321
5	0.6153
6	0.6259
7	0.6238
8	0.6245
9	0.6270
10	0.6265
11	0.6264
12	0.6270

EXAMPLE. *Censored Regression (Continued)*

As noted in the example of Section 4.1, the log-augmented posterior under a flat prior is (up to a constant)

$$-n \log \sigma - \sum_{j=1}^{k}(t_j - \beta_0 - \beta_1 v_j)^2/2\sigma^2 - \sum_{j=k+1}^{n}(Z_j - \beta_0 - \beta_1 v_j)^2/2\sigma^2.$$

The conditional predictive distribution $p(Z_j \mid \beta_0^i, \beta_1^i, \sigma_i, c_j)$ is the conditional normal distribution, conditional on the fact that the unobserved failure time Z_j is greater than the observed right-censored time c_j. That is, sample from

$$\frac{\phi(x)}{1 - \Phi\left(\frac{c_j - \beta_0^i - \beta_1^i v_j}{\sigma_i}\right)}, \quad \frac{c_j - \beta_0^i - \beta_1^i v_j}{\sigma_i} < x < \infty,$$

where $\phi(x)$ and $\Phi(x)$ are the density and cdf of the standard normal distribution, respectively, and β_0^i, β_1^i and σ_i are the current estimates of the intercept, slope and scale parameter, respectively.

A failure time is imputed by sampling from the appropriate conditional normal distribution. Thus, the integration $\int_c^{\infty} x\phi(x)dx$ in *EM* is replaced by random draws from the conditional normal in Monte Carlo *EM*.

Having imputed failure times for $n - k$ censored observations to obtain an augmented data set $\underset{\sim}{\mu}_1$, the process is repeated to obtain m augmented data sets. It follows that in the $\tilde{M}$-step

$$\underset{\sim}{\beta}^{i+1} = (V^T V)^{-1} V^T \left(\sum_{j=1}^{m} \underset{\sim}{\mu}_j/m\right)$$

TABLE 4.3.   History of *MCEM*

Iteration	β_0	β_1	$\sigma^2(\times 100)$
1	−5.27	3.93	3.36
2	−5.61	4.10	4.07
3	−5.64	4.12	4.65
4	−5.77	4.18	5.08
5	−5.81	4.20	5.31
6	−5.84	4.22	5.41
7	−5.85	4.23	5.62
8	−5.86	4.23	5.70
9	−5.94	4.27	5.72
10	−5.88	4.24	5.82
11	−5.97	4.29	5.93
12	−5.88	4.24	5.80
13	−5.78	4.19	5.77
14	−5.94	4.27	5.67
15	−5.97	4.28	5.84
16	−5.96	4.28	5.89
17	−5.96	4.28	5.89
18	−5.96	4.28	5.89

and

$$\sigma_{i+1}^2 = \sum_{j=1}^{m} (\underset{\sim}{\mu_j} - V\underset{\sim}{\beta^{i+1}})^T (\underset{\sim}{\mu_j} - V\underset{\sim}{\beta^{i+1}})/mn,$$

where V is the $n \times 2$ design matrix with 1's in the first column and the v_j's in the second column.

For the Schmee and Hahn (1979) data, the algorithm was initiated with $\beta_0^0 = -4.931$, $\beta_1^0 = 3.747$ and $\sigma^0 = \sqrt{0.0247}$. The value of m was equal to 50 (5000) for iterations 1–14 (15–18). The values $\beta_0 = -6.02$, $\beta_1 = 4.31$ and $\sigma^2 = 0.067$ from the final three iterations agree with the values quoted in the example of Section 4.1.

Table 4.3 gives history of the Monte Carlo *EM* algorithm implemented under the noninformative prior $p(\beta, \sigma^2) \propto 1/\sigma^2$. The same starting point and values of m were used as in the case given above.

Sinha, Tanner and Hall (1994) apply *MCEM* to address the analysis of grouped survival data via a marginal likelihood approach by imputing ranks. Guo and Thompson (1991; 1994) use *MCEM* to analyze variance component models on complex pedigrees.

4.6 Acceleration of *EM* (Louis' Turbo *EM*)

As noted in Dempster, Laird and Rubin (1977), the convergence rate of the *EM* algorithm is linear and is governed by the fraction of missing information. Thus, when the proportion of missing data is high, convergence can be quite slow. In this regard, Louis (1982) has proposed a device for accelerating the convergence of the *EM* algorithm.

Recall the Newton–Raphson iteration:

$$\theta^{m+1} = \theta^m + \left[\frac{-\partial^2 \ \log \ p(\theta \mid Y)}{\partial \theta^2} \bigg|_{\theta*} \right]^{-1} \frac{\partial \ \log \ p(\theta \mid Y)}{\partial \theta} \bigg|_{\theta*}$$

Newton–Raphson has a quadratic convergence rate in a neighborhood of the mode. The goal is to be able to use *EM* "type" quantities to perform a Newton–Raphson step, i.e. achieve quadratic convergence near the mode.

Let θ_{EM}^m denote the update to the posterior mode when the *EM* algorithm is started at θ^m. Expanding $\partial \ Q(\theta, \theta^m)/\partial \theta$ about θ^m, we have

$$0 = \frac{\partial \ Q(\theta, \theta^m)}{\partial \theta} \bigg|_{\theta_{EM}^m} \approx \frac{\partial \ Q(\theta, \theta^m)}{\partial \theta} \bigg|_{\theta^m} + \left[\frac{\partial^2 Q(\theta, \theta^m)}{\partial \theta^2} \bigg|_{\theta^m} \right] (\theta_{EM}^m - \theta^m).$$

Hence, in the limit we have

Result 4.6.1.

$$\frac{\partial \ Q(\theta, \theta^m)}{\partial \theta} \bigg|_{\theta^m} = \frac{-\partial^2 \ Q(\theta, \theta^m)}{\partial \theta^2} \bigg|_{\theta^m} (\theta_{EM}^m - \theta^m).$$

As noted in Section 4.4.3, we have

Result 4.6.2.

$$\frac{\partial \ \log \ p(\theta \mid Y)}{\partial \theta} = \int \frac{\partial \ \log \ p(\theta \mid Y, Z)}{\partial \theta} p(Z \mid Y, \theta) dZ = \frac{\partial \ Q(\theta, \theta')}{\partial \theta} \bigg|_{\theta'=\theta}.$$

Combining Results 4.6.1 and 4.6.2 with the Newton–Raphson step, we have

$$\theta^{m+1} = \theta^m + \left[\frac{-\partial^2 \ \log \ p(\theta \mid Y)}{\partial \theta^2} \bigg|_{\theta^m} \right]^{-1}$$

$$\times \left[-\int \frac{\partial^2 \ \log \ p(\theta \mid Y, Z)}{\partial \theta^2} p(Z \mid Y, \theta^m) dZ \bigg|_{\theta^m} \right] (\theta_{EM}^m - \theta^m).$$

Louis (1982) points out that this formula is an example of Aitken's acceleration method. Louis also notes that this approximation is the most useful in a neighborhood of the mode. In this regard, further experience is required to develop rules about when to perform such an update.

Note that $[-\partial^2 \log p(\theta \mid Y)/\partial\theta^2] \mid_{\theta^m}$ may be replaced by

$$-\int \frac{\partial^2 \log p(\theta \mid Y, Z)}{\partial\theta^2} p(Z \mid \theta^m, Y)dZ \bigg|_{\theta^m} - \text{var}\left[\frac{\partial \log p(\theta \mid Y, Z)}{\partial\theta} \bigg|_{\theta^m} \right].$$

Finally, it is noted that the Monte Carlo *EM* algorithm can be accelerated in an analogous manner. Jamshidian and Jennrich (1993) discuss an approach to accelerating the *EM* algorithm based on conjugate gradients. See also Lange (1995a, 1995b).

EXAMPLE. *Genetic Linkage (Continued)*
To illustrate the Louis' Turbo *EM* algorithm, initiate the *EM* algorithm at $\theta^0 = 0.6$ to obtain

$$\theta^1 = 0.623188$$
$$\theta^2 = 0.626338$$
$$\theta^3 = 0.626757$$
$$\theta^3_{EM} = 0.626812$$
$$\theta^* = 0.6268215 \ldots$$

In this case,

$$[-\partial^2 \log p(\theta \mid Y)/\partial\theta^2 \mid_{\theta^3}]^{-1} \left[-\int \frac{\partial^2 \log p(\theta \mid Y, Z)}{\partial\theta^2} p(Z \mid Y, \theta^3)dZ \bigg|_{\theta^3} \right]$$

$$= \frac{434.79}{376.95} = 1.153442.$$

It then follows that

$$\theta^4 = 0.626757 + 1.153442(0.626812 - 0.626757) = 0.6268205.$$

4.7 Facilitating the M-Step

The *EM* algorithm is less attractive when the M-step is computationally intensive. Meng and Rubin (1993) note that is may be advantageous to replace a complicated M-step with several computationally simpler conditional maximization steps.

More specifically, the *ECM* (Expectation/Conditional Maximization) algorithm replaces each M-step of *EM* by a sequence of S conditional maximization steps, each of which maximizes the Q function over the parameter space θ but for some vector function of θ, $h_s(\theta)(s = 1, \ldots, S)$ fixed at its previous value. For example, Meng and Rubin (1993) in one particular setting find it useful to partition θ into a location parameter (θ_1) and a scale parameter (θ_2). An *ECM* algorithm would be to maximize the Q function over θ_1 holding θ_2 fixed and then maximize the Q function over θ_2 given the new value for θ_1. Meng and Rubin (1993) show that any *ECM* algorithm is a *GEM* algorithm, and they present conditions under which the *ECM* algorithm converges to the observed likelihood or posterior mode. Extensions and further work are presented in Liu and Rubin (1994, 1995) and Meng (1994b).

Exercises

1. Recall the genetic linkage model of Section 4.1.
   a. For the data $Y = (125, 18, 20, 34)$ implement the *EM* algorithm. Use a flat prior on θ. Try starting your algorithm at $\theta = .1, .2, .3, .4, .6$ and .8. Did the algorithm converge for all of these starting values? How did you assess convergence? How many iterations were required for convergence?
   b. Compute the standard error via Louis' method. Compute the standard error via the simulation method of Section 4.4.4. Compare your results obtained via these two methods.
   c. Plot the normal approximation along with the normalized likelihood. Is the normal approximation appropriate in this case?
   d. Repeat (a)–(c) for the data $Y = (14, 0, 1, 5)$. Did the algorithm converge for all of the above starting values? Compare to Exercise 9f of Chapter 2.
2. Repeat Exercise 1 using the Monte Carlo *EM* algorithm. How did you assess convergence? What value(s) of m did you use?
3. Let $x_1, \ldots x_n$ be an independent sample from a population such that the distribution of x_i given μ, σ and c_i is $N(\mu, \sigma^2/c_i)$, where c_i are unobserved *iid* positive scalar random variables with known density $f(c_i)$. We want to use the *EM* algorithm to obtain the likelihood mode $(\hat{\mu}, \hat{\sigma}^2)$ treating the c_i as missing data (Little and Rubin, 1987).
   a. Derive the Q function for this problem.
   b. Let μ' and σ_t^2 be the current value for the mode. Show that the E-step reduces to computing $w_i' = E(c_i \mid x_i, \mu', \sigma_t^2)$.
   c. Show that the M-step reduces to computing

   $$\mu^{t+1} = \sum_{i=1}^{n} w_i' x_i / \sum_{i=1}^{n} w_i',$$

   $$\sigma_{t+1}^2 = \sum_{i=1}^{n} \frac{w_i'(x_i - \mu')^2}{n},$$

   i.e. the *EM* algorithm is a form of iteratively reweighted least squares.
   d. When c_i is distributed as χ_ν^2/ν, what is the unconditional distribution of x_i? What is w_i' in this case?
   e. Apply this algorithm to the Darwin data of Exercise 2 of Chapter 2, using the above distribution for c_i with $\nu = 4$. How did you assess convergence? How many iterations were required for convergence? Try several randomly selected starting points. Is there evidence that the observed likelihood is multimodal? Compare to the results of Exercise 5 of Chapter 3.
4. Let $t_1, \ldots, t_n$ be *iid* with density $\rho e^{-\rho t}$, i.e. the exponential distribution with

parameter ρ. Suppose that we observe (y_i, δ_i), where $y_i = \min(t_i, c_i)$ and $\delta_i = 1$ if $t_i < c_i$ and 0 otherwise, for $i = 1, \ldots, n$.

a. Show that the Q function (under a flat prior for ρ) is $n \log \rho - \rho \sum_{i=1}^{n} E(t_i \mid y_i, \delta_i, \rho')$, where ρ' is the current approximation to the mode.

b. For an uncensored observation, we have $t_i = y_i$. For a censored observation, what is $E(t_i \mid y_i, \delta_i, \rho')$?

c. What is the M-step for this problem?

d. Implement the EM algorithm using the Stanford Heart Transplant data listed in Chapter 1 (ignore the covariates in this exercise). How did you assess convergence? How many iterations were required for convergence? Try several randomly selected starting values. Is there evidence that the observed likelihood is multimodal?

e. For the Stanford Heart Transplant data, compute the standard error via Louis' method. Compute the standard error via the simulation method of Section 4.4.4. Compare your results obtained via these two methods.

f. For the Stanford Heart Transplant data, plot the normal approximation along with the normalized likelihood. Is the normal approximation appropriate in this case?

g. For the Stanford Heart Transplant data, locate the mode using Newton–Raphson. Compare these results to those in parts (d) and (e).

5. Derive the conditional expectation of $\alpha^T \alpha$, $\varepsilon^T \varepsilon$ and $\mathbf{1}^T (y - Z\alpha)$ for the variance components example in Section 4.3.

6. a. Use the EM algorithm to apply the simple linear regression model of Section 4.1 to the Stanford Heart Transplant data using age as the covariate. How did you assess convergence? How many iterations were required for convergence? Try several randomly selected starting points. Is there evidence of multimodality to the posterior distribution?

b. Obtain the standard error of the slope and intercept using Louis' method.

c. Repeat (a) using the Monte Carlo EM algorithm. What value(s) did you take for m?

d. Repeat (a) and (b) using the mismatch score as the covariate. How did you handle the missing values for this covariate?

7. Consider the k-component mixture model for the observations $x_1, \ldots, x_n$:

$$f(x_i) = \sum_{j=1}^{k} p_j \phi(x_i; \mu_j; \sigma),$$

for $i = 1, \ldots, n$ where

$$\phi(x_i; \mu_j; \sigma) = \frac{1}{\sqrt{2\pi}\sigma} \exp\left(-\frac{(x_i - \mu_j)^2}{2\sigma^2}\right),$$

the μ_j, p_j and σ are unknown, $\sum_{i=1}^{k} p_j = 1$ and $0 < p_j < 1$. For $1 \leq i \leq n$, let z_i denote a k-dimensional vector indicating to which component x_i belongs, such that $z_{ij} \in \{0, 1\}$ and $\sum_{j=1}^{k} z_{ij} = 1$. The augmented likelihood (Diebolt and Robert, 1994) is then

$$\prod_{i=1}^{n} \prod_{j=1}^{k} p_j^{z_{ij}} (\phi(x_i; \mu_j; \sigma))^{z_{ij}}.$$

a. Derive the Q function for this problem. What is the E-step? What is the M-step?

b. The following table represents the velocities at which 82 galaxies in the Corona Borealis region are moving away from our galaxy (Roeder, 1990). If the galaxies are clustered, then the density of the velocities will be multimodal. Fit the above model to these data using the *EM* algorithm. Take $k = 3$. How did you assess convergence? How many iterations were required for convergence? Try several randomly selected starting points for the algorithm. Is there evidence that the observed likelihood is multimodal?

Velocities of Galaxies for an Unfilled
Survey of the Corona Borealis Region
Velocity (km per second)

9,172	9,350	9,483
9,558	9,775	10,227
10,406	16,084	16,170
18,419	18,552	18,600
18,927	19,052	19,070
19,330	19,343	19,349
19,440	19,473	19,529
19,541	19,547	19,663
19,846	19,856	19,863
19,914	19,918	19,973
19,989	20,166	20,175
20,179	20,196	20,215
20,221	20,415	20,629
20,795	20,821	20,846
20,875	20,986	21,137
21,492	21,701	21,814
21,921	21,960	22,185
22,209	22,242	22,249
22,314	22,374	22,495
22,746	22,747	22,888
22,914	23,206	23,241
23,263	23,484	23,538
23,542	23,666	23,706
23,711	24,129	24,285
24,289	24,366	24,717

Velocities of Galaxies for an Unfilled
Survey of the Corona Borealis Region
Velocity (km per second) *(continued)*

24,990	25,633	26,960
26,995	32,065	32,789
34,279		

c. Obtain standard errors using Louis' method.
d. Fit the above mixture model to these data using Newton–Raphson. Compare these results to those in (b) and (c).

8. Suppose that from a population of unknown size N, we draw, mark and return to the population I samples of sizes $n_1, \ldots n_I$. Let the total number of distinct captured individuals be denoted and recorded as r. Note that $r = \Sigma_{i=1}^{I}(n_i - m_i)$, where m_i have been captured previously. Suppose that the probability that individual j is captured in sample i is ρ_i. If the $n_1, \ldots, n_I$ are treated as random, then the augmented likelihood (treating N as the missing data) is

$$\frac{N!}{(N-r)!} \prod_{i=1}^{I} \rho_i^{n_i}(1-\rho_i)^{N-n_i}$$

a. Show that under the prior $p(N, \rho) = p(N)p(\rho)$, where $\rho = (\rho_1, \ldots, \rho_I)$, the conditional distribution of N given ρ and the data is proportional to

$$\frac{N!}{(N-r)!}\left(\prod_{i=1}^{I}(1-\rho_i)\right)^{N} p(N).$$

b. Show that if $p(N) = 1/N$, i.e. the noninformative prior, then the distribution in (a) is negative binomial with parameters $r-1$ and $1-\prod_{i=1}^{I}(1-\rho_i)$.
c. Derive the Q function for this problem. What is the E-step? What is the M-step?
d. Apply this model using the *EM* algorithm to the following data (George and Robert, 1992). How did you assess convergence? How many iterations were required for convergence? Try several randomly selected starting values. Is there evidence that the observed likelihood is multimodal?
e. Use Louis' method to obtain standard errors.

					Capture–Recapture Counts of Gordy Lake Sunfish									
i	1	2	3	4	5	6	7	8	9	10	11	12	13	14
n_i	10	27	17	7	1	5	6	15	9	18	16	5	7	19
m_i	0	0	0	0	0	0	2	1	5	5	4	2	2	3

5

The Data Augmentation Algorithm

5.1 Introduction and Motivation

Analogous to the *EM* algorithm, the data augmentation algorithm exploits the
simplicity of the likelihood function or posterior distribution of the parameter given
the augmented data. In contrast to the *EM* algorithm, the present goal is to obtain the
entire (normalized) likelihood or posterior distribution, not just the maximizer and
the curvature at the maximizer. In large samples, it is comforting that the posterior
or likelihood is consistent with the normal approximation, though in practice it is
not often clear when one is in a large sample setting. In a small sample situation,
the data augmentation algorithm will provide a way of improving the inference,
based on the entire posterior distribution or the entire likelihood function.

The basic idea behind the data augmentation algorithm is to augment the *observed data* Y by a quantity Z, which will be referred to as *latent data*. It is assumed
that given both Y and Z, one can calculate or sample from the augmented posterior $p(\theta \mid Y, Z)$. To obtain the posterior $p(\theta \mid Y)$, one generates multiple values
(imputations) of Z from the predictive distribution $p(Z \mid Y)$ and then computes
the average of $p(\theta \mid Y, Z)$ over the imputations. Because $p(Z \mid Y)$ depends on
$p(\theta \mid Y)$, one obtains an iterative algorithm to calculate $p(\theta \mid Y)$.

The data augmentation algorithm is motivated by two simple identities. The
posterior identity is given by

$$p(\theta \mid Y) = \int_Z p(\theta \mid Y, Z) p(Z \mid Y) dZ,$$

where $p(\theta \mid Y)$ denotes the posterior density of the parameter θ given the observation Y, $p(Z \mid Y)$ denotes the predictive density of the latent data Z given Y,
and $p(\theta \mid Y, Z)$ denotes the conditional density of θ given the augmented data
$X = (Y, Z)$, i.e. the augmented posterior. The *predictive identity* is given by

$$p(Z \mid Y) = \int_\Theta p(Z \mid \phi, Y) p(\phi \mid Y) d\phi,$$

where $p(Z \mid \phi, Y)$ is the conditional predictive distribution. Substituting the predictive identity into the posterior identity and interchanging the order of integration,

90

$p(\theta \mid Y)$ satisfies the integral equation

$$g(\theta) = \int K(\theta, \phi)g(\phi)d\phi,$$

where

$$K(\theta, \phi) = \int p(\theta \mid Z, Y)p(Z \mid \phi, Y)dZ \qquad (5.1.1)$$

and $g(\theta) = p(\theta \mid Y)$. The method of successive substitution can now be used to solve (5.1.1). In other words, starting with an initial approximation $g_0(\theta)$, calculate successively

$$g_{i+1}(\theta) = (Tg_i)(\theta), \qquad (5.1.2)$$

where

$$Tf(\theta) = \int K(\theta, \phi)f(\phi)d\phi.$$

Tanner and Wong (1987) adopt the method of Monte Carlo to perform the integration in (5.1.1). In particular, applying the method of Monte Carlo to the posterior identity yields the following iterative scheme:

a. Generate a sample $z_1, \ldots, z_m$ from the current approximation to the predictive distribution $p(Z \mid Y)$.
b. Update the current approximation to $p(\theta \mid Y)$ to be the mixture of augmented posteriors of θ, given the augmented data from (a), i.e.

$$g_{i+1}(\theta) = \frac{1}{m} \sum_{j=1}^{m} p(\theta \mid z_j, Y).$$

Given the current guess to the posterior, $(g_i(\theta))$, the method of composition is applied to the predictive identity to generate a sample of latent data.

a1. Generate θ^* form $g_i(\theta)$.
a2. Generate z from $p(Z \mid \theta^*, Y)$, where θ^* is the value generated in (a1).

These two steps are then repeated m times to obtain $z_1, \ldots, z_m$. Rubin (1987b) refers to the quantities $z_1, \ldots, z_m$ as *multiple imputations*. In this regard, we call step (a) the *imputation step*, while step (b) is referred to as the *posterior step*. The data augmentation algorithm consists of iterating between the imputation and the posterior steps.

Two important practical considerations in the implementation of the data augmentation algorithm are the monitoring of convergence and the selection of the number of imputations (m) to be performed per iteration. When m is large, steps (a) and (b) will provide a close approximation to the iteration of (5.1.2). Further comments are presented in Section 5.4. Regularity conditions under which the algorithm converges are given in Section 5.5.

EXAMPLE. *Genetic Linkage (Continued)*
As noted in Chapter 4, the augmented posterior for the genetic linkage model is
of the beta form

$$p(\theta \mid Y, Z) \propto \theta^{x_2+x_5}(1 - \theta)^{x_3+x_4}.$$

The conditional predictive distribution $p(Z \mid \theta, Y)$ is the binomial distribution
with $n = 125$ and $p = \theta/(\theta + 2)$. Thus, the algorithm is given as

a. Imputation Step

a1. Draw θ^* from the current estimate of the posterior.
a2. Generate x_2 by drawing from the binomial distribution with parameters
   $[125, \theta^*/(\theta^* + 2)]$.

Repeat steps (a1) and (a2) m times to yield $x_2^{(1)} \ldots x_2^{(m)}$.

b. Posterior Step
Set the posterior density of θ equal to the mixture of beta distributions, mixed over
the m imputed values of x_2; i.e.

$$p(\theta \mid Y) = \frac{1}{m} \sum_{i=1}^{m} \text{Beta}(v_1^{(i)}, v_2^{(i)})(\theta),$$

where

$$v_1^{(i)} = x_2^{(i)} + x_5 + 1, \qquad v_2^{(i)} = x_3 + x_4 + 1$$

and

$$\text{Beta}(v_1, v_2)(\theta) = \frac{\Gamma(v_1 + v_2)}{\Gamma(v_1)\Gamma(v_2)} \theta^{v_1-1}(1 - \theta)^{v_2-1}.$$

In this example, the prior for θ is assumed to be uniform in $(0,1)$. Steps (a1), (a2)
and (b) are to be iterated until convergence of the algorithm is achieved.
   In step (a1) of the first iteration, one would draw a value for θ from a reasonable
starting distribution (e.g. uniform). This value for θ would then be used to generate
a binomial deviate in step (a2). Steps (a1) and (a2) would then be repeated m times
to generate m imputations. In subsequent iterations, the current approximation to
the posterior is a mixture of beta distributions. To sample a value for θ from the
mixture of betas

$$\frac{1}{m} \sum_{j=1}^{m} \text{Beta}(v_1^{(i)}, v_2^{(i)})(\theta),$$

one randomly selects a distribution (e.g. Beta$(v_1^{(j)}, v_2^{(j)})(\theta)$) from this mixture and
then draws a θ value from this beta distribution.
   Figure 5.1 presents the posterior density estimates of θ for this example. In
particular, the normal approximation with $\hat{\mu} = 0.63$ and $\hat{\sigma} = 0.05$ (solid line) is
plotted along with the true posterior distribution (dotted line)

$$p(\theta \mid Y) \propto (2 + \theta)^{y_1}(1 - \theta)^{y_2+y_3}\theta^{y_4}$$

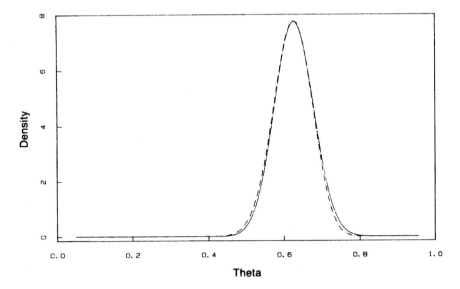

FIGURE 5.1. Posterior distribution of θ.

and the estimated posterior (dashed line) obtained by plotting the mixture of the beta distributions at the final iteration in which $m = 1600$. In the density scale, all three estimates are congruent. If possible, it is preferable to use the mixture as the approximation to the posterior rather than using the histogram of the generated θ values. However, as in the next example, this may not always be possible, though the approach due to Chen (1994)—see Section 5.9.3—may help.

Alternatively, a second version of these data is examined in which the sample size is reduced by a factor of 10, though the cell proportions are approximately unchanged; i.e. $Y = (13, 2, 2, 3)$. The resulting posterior density estimates are plotted in Figure 5.2. In this case, although the true posterior density and the estimated posterior density are congruent, the validity of the normal approximation may be in doubt, even when viewed on the density scale.

An even more dramatic illustration is given in Figure 5.3, where $Y = (14, 0, 1, 5)$. In cases with such a dramatic departure from normality, one or two iterations of the data augmentation algorithm indicate the inadequacy of the normal approximation.

EXAMPLE. *Bivariate Normal Covariance Matrix*
Suppose that the data in Table 5.1 represent 12 observations from the bivariate normal distribution with $\mu_1 = \mu_2 = 0$, correlation coefficient ρ and variances σ_1^2 and σ_2^2. Note that in the four pairs of observations, two pairs have correlation 1 and

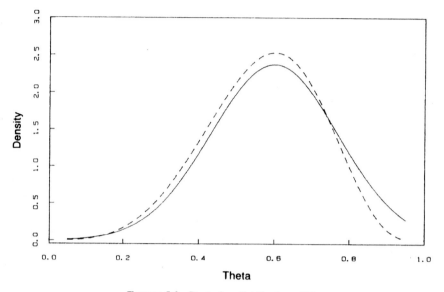

FIGURE 5.2. Posterior distribution of θ.

FIGURE 5.3. Posterior distribution of θ.

TABLE 5.1.   Twelve Observations from a Bivariate Normal Distribution.

1	1	−1	−1	2	2	−2	−2	*	*	*	*
1	−1	1	−1	*	*	*	*	2	2	−2	−2

*Value not observed (missing at random).

the remaining two pairs have correlation −1. Thus, one can expect a nonunimodal posterior distribution for ρ in this data set. In such a case, the maximum likelihood estimate and the associated standard error will clearly be misleading. Furthermore, note that the information regarding σ_1^2 and σ_2^2 in the eight incomplete observations cannot be ignored because information regarding σ_1^2 and σ_2^2 is of use in making an inference regarding ρ.

In this example, the conditional predictive distribution $(p(Z \mid \theta, Y))$ is the conditional normal. Under the prior on Σ given as

$$p\left(\sum\right) \propto \left|\sum\right|^{-(d+1)/2},$$

where d is the dimension of the observation vector, the augmented posterior of Σ is an inverted Wishart distribution (Box and Tiao, 1973). Thus, the data augmentation algorithm is as given below.

a. Imputation Step

a1. Draw Σ from the current estimate of the observed posterior.
a2. Generate latent data as follows:

If x_1 is known, then generate the unobserved observation from

$$N\left[\rho\frac{\sigma_2}{\sigma_1}x_1, \sigma_2^2(1-\rho^2)\right].$$

If x_2 is known, then generate the unobserved observation from

$$N\left[\rho\frac{\sigma_1}{\sigma_2}x_2, \sigma_1^2(1-\rho^2)\right].$$

Repeat steps (a1) and (a2) m times.

b. Posterior Step
Set the posterior density of Σ equal to the mixture of inverted Wishart distributions.

Regarding the implementation of the algorithm, it is noted that the algorithm of Odell and Feiveson (1966) can be used to generate observations from the inverted Wishart distribution. The amount of computation in this algorithm is not extensive, since the computation is of order $d(d+1)/2$, which does not depend on the sample size.

In Figure 5.4, the histogram of the imputed correlation coefficients based on pooling iterations 10 to 15 ($m = 6400$) is presented. In addition, the true posterior of the correlation coefficient, which is proportional to $(1 - \rho^2)^{4.5}/(1.25 - \rho^2)^8$, is also plotted. As is evident from the plot, the estimated posterior distribution recovers the bimodal nature of the true distribution.

Finally, it is noted that the data augmentation algorithm can be used to examine the posterior distribution of any functional of the covariance matrix. For example, the posterior distribution of the smallest eigenvalue of the covariance matrix (Tiao and Fienberg, 1969) may be examined by simply computing the smallest eigenvalue of each of the observations sampled from the mixture of inverted Wishart distributions computed in step (b) of the algorithm (see Figure 5.5).

EXAMPLE. *Censored Regression (Continued)*
To apply the data augmentation algorithm, we need to know how to sample from an augmented posterior and how to sample from the conditional predictive distribution. As in Section 4.1, we assume that the errors follow the standard normal distribution. See Wei and Tanner (1990c).

In the case of no censoring, the posterior density function of $(\sigma^2, \beta_0, \beta_1)$ can be factored exactly into the product of the marginal posterior density of σ^2 and the conditional marginal density of (β_0, β_1) given σ^2. In fact, for the noninformative

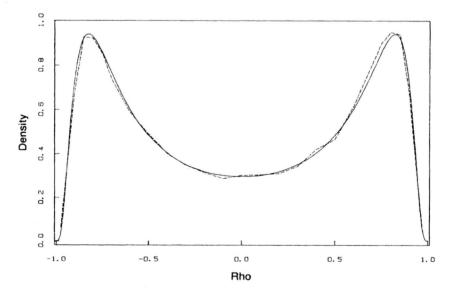

FIGURE 5.4. Posterior distribution of ρ.

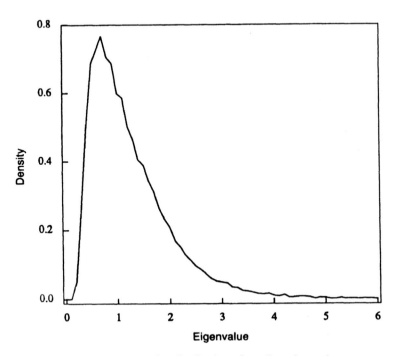

FIGURE 5.5. Posterior distribution of smallest eigenvalue.

prior, it is well known (Box and Tiao, 1973) that the marginal posterior distribution of σ^2 is that of the distribution of the random variable $(n-d)s^2/\chi^2_{n-d}$, where χ^2_v is a chi-square random variable with degrees of freedom v. The conditional marginal posterior distribution of (β_0, β_1) is a conditional bivariate normal distribution centered at the least-squares estimate of (β_0, β_1) for the given data set, conditional on σ^2.

The factorization in the previous paragraph is a convenient tool for the sampling of $(\sigma^2, \beta_0, \beta_1)$ from an augmented posterior. In particular, for an augmented data set, one first generates σ^2_* from $p(\sigma^2 \mid s^2)$, then generates a value of (β_0, β_1) from the conditional bivariate normal distribution.

To impute a failure time for the right-censored observation given the values σ_* and (β_0, β_1), one simply draws from the conditional distribution of the error, conditional on the fact that the failure time must be larger than the observed event time (see Figure 5.6). Had the data been left censored, to impute a failure time one would draw from the conditional error distribution, conditional on the fact that the failure time must be *smaller* than the observed event time. Had the data been interval censored, to impute a failure time, one would draw from the conditional

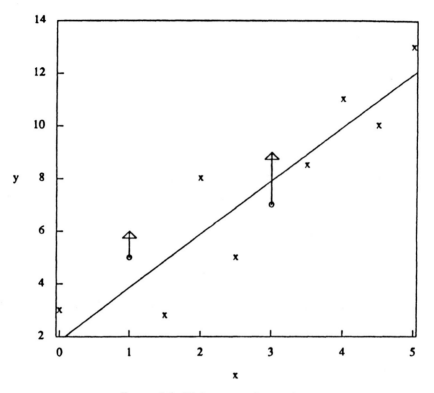

FIGURE 5.6. Right-censored event data.

error distribution, conditional on the fact that the failure must have occurred in a particular interval.

In the present situation of right-censored normal data, we draw from the distribution

$$\frac{\phi(s)}{1 - \Phi(z_i)}, \qquad z_i < s < \infty,$$

where $\phi(s)$ and $\Phi(s)$ are the density and cdf of the standard normal distribution, respectively, and z_i is the value

$$z_i = \frac{c_i - \beta_0^* - \beta_1^* x_i}{\sigma_*}.$$

Having presented an approach for generating the tuple $(\sigma, \beta_0, \beta_1)$ from an augmented posterior, as well as for generating the latent data, the data augmentation algorithm for the regression analysis of censored data with normal errors is given below.

a1. Generate (β_0, β_1) and σ^2 from the current guess to the posterior.
a2. Generate latent data from the conditional normal distributions.

Repeat steps (a1) and (a2) m times.

b. Update the posterior as a mixture of augmented data posteriors, mixed over the latent data, generated in part (a) of the algorithm.

In general, one may stop the data augmentation algorithm at any point and realize a sample from the exact posterior distribution $p(\theta \mid Y)$, rather than from the current approximation $g_i(\theta)$. In particular, by attaching weights (w_j) proportional to

$$\frac{p(\theta_j \mid Y)}{g_i(\theta_j)}$$

to the sample values $\theta_1, \theta_2, \ldots, \theta_n$ from $g_i(\theta)$, one obtains a sample from the desired distribution. (Note that one needs to know $p(\theta \mid Y)$ only up to the normalizing constant.) However, as pointed out in the rejoinder in Tanner and Wong (1987), because the variance of any function of these θ and weight values will depend on the values

$$\frac{p(\theta_j \mid Y)^2}{g_i(\theta_j)},$$

it is important to iterate the data augmentation algorithm to help eliminate outliers that would inflate the (Monte Carlo) variance of the estimate. Section 5.4 discusses how the w_j can be used to assess convergence of the algorithm.

Figure 5.7 presents the posterior density of β_1 for the censored regression problem calculated via the data augmentation algorithm (solid and long-dash lines), along with the normal approximation based on the maximum likelihood solution (dotted line) and the true posterior density (short-dash line). The true posterior was calculated by numerical integration. A histogram of the 300 values of β_1 from the 12–15 iterations ($m = 100$) is given by the solid curve. The corresponding curve for the weighted values (weighted according to p/g) is given by the long-dash curve. As can be seen, the normal approximation did not capture the slight skew of the true posterior. Both the weighted and unweighted data augmentation curves detected this skew. If a more precise estimate is required, then the value of m could be increased and the algorithm iterated further.

EXAMPLE. *Tobit Model*
Chib (1992) considers the Tobit model, which has some similarities to the censored regression model. In the Tobit model

$$t_i = x_i^T \beta + \varepsilon_i,$$

for $i = 1, \ldots, n$, where β is the $d \times 1$ regression vector, $x_1, \ldots, x_n$ are $d \times 1$ vectors of known constants and $\varepsilon_1, \ldots, \varepsilon_n$ is an *iid* sample from $N(0, \sigma^2)$ with

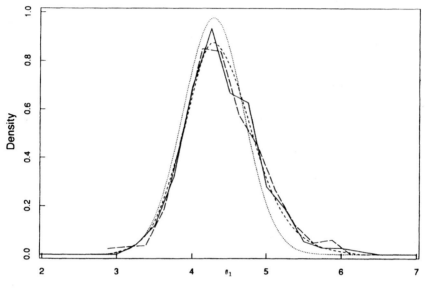

FIGURE 5.7. Posterior distribution of β.

σ unknown. We observe $y_i = \max(t_i, 0)$ rather than the minimum of a failure time and a censoring time as was the case in the previous example. For the Tobit censored observations, Chib (1992) imputes values from $p(Z \mid \beta, \sigma)$ by sampling from the conditional normal with mean $x_i^T \beta$ and variance σ^2 *conditional on the fact that the random variable Z is less than* 0. Having imputed a value for each of the Tobit censored observations, the augmented posterior $p(\beta, \sigma \mid Y, Z)$, like the previous censored regression example, has a multivariate (conditional) normal by inverse chi-square form (taking a flat prior on β and a noninformative prior on σ).

EXAMPLE. *Dichotomous Response Data*
Albert and Chib (1993) consider the Bayesian analysis of binary response data. They consider the situation of N independent binary random variables $Y_1, \ldots, Y_n$, where Y_i is distributed as a Bernoulli random variable with probability of success $p_i = P(Y_i = 1) = \Phi(x_i^T \beta)$. Here β is a $d \times 1$ vector of unknown parameters, x_i is a $d \times 1$ vector of known covariates, and $\Phi(.)$ is the standard normal cdf. Albert and Chib (1993) introduce n independent latent variables $Z_1, \ldots, Z_n$, where Z_i is normally distributed with mean $x_i^T \beta$ and variance unity. They let $Y_i = 1$ if $Z_i > 0$ and $Y_i = 0$ if $Z_i \leq 0$. In other words, if the latent variable crosses the threshold, then a success is observed, otherwise a failure is observed. It then follows that the

Y_i are independent Bernoulli random variables with parameters $p_i = P(Y_i = 1)$ $= \Phi(x_i^T \beta)$.

To implement data augmentation in this case, Albert and Chib (1993) note that when $Y_i = 1$, the conditional predictive distribution $p(Z \mid Y, \beta)$ is $N(x_i^T \beta, 1)$ truncated on the *left* by 0. When $Y_i = 0$, $p(Z \mid Y, \beta) = N(x_i^T \beta, 1)$ truncated on the *right* by 0. Note that from this point of view all the responses are "missing." The augmented posterior $p(\beta \mid Y, Z)$ (under a flat prior for β) is normal with mean $\hat{\beta} = (X^T X)^{-1} X^T Z$ and variance $(X^T X)^{-1}$, where $X^T = (x_1, \ldots, x_n)$ and $Z^T = (Z_1, \ldots, Z_n)$. (The posterior distribution is not the conditional normal by chi-square in this case, since σ is specified to equal unity.) Albert and Chib (1993) also consider a hierarchical version of the model and generalizations to multinomial responses. Albert and Chib (1995) discuss Bayesian residual analysis for binary response regression models. Cowles, Carlin and Connett (1996) consider a longitudinal trivariate categorical Tobit model for assessing compliance in a clinical trial. See also Zellner and Rossi (1984).

5.2 Computing and Sampling from the Predictive Distribution

Having obtained a sample from $p(\theta \mid Y)$, calculating the posterior distribution of any functional of θ is straightforward. In this section, we focus on the predictive distribution for a future observation. In the following section, we consider the content and boundary of the highest posterior density region.

Suppose y_f is a (univariate) future observation from the model

$$y_f = v_f^T \beta + \sigma \varepsilon_f.$$

As noted in Section 3.3.2, the predictive distribution for y_f is given by

$$p(y_f \mid Y) = \int \phi(y_f \mid v_f^T \beta, \sigma^2) p(\sigma^2, \beta \mid Y) d\sigma^2 d\beta.$$

When the errors follow the normal distribution, $\phi(x \mid \mu, \sigma^2)$ is the density of the normal distribution.

Following the Monte Carlo method, given a sample of size m from $p(\sigma^2, \beta \mid Y)$, one may evaluate the expression

$$\frac{1}{m} \sum_{i=1}^{m} \phi(y_f \mid v_f^T \beta_i, \sigma_i^2)$$

over the m β and σ^2 values to obtain an estimate of the predictive distribution.

One may sample from $p(y_f \mid Y)$ by first drawing (σ_0^2, β_0) from $p(\sigma^2, \beta \mid Y) d\sigma^2 d\beta$ and then generating a random number from $\phi(x \mid \sigma_0^2, \beta_0)$. Repeating this two-stage algorithm yields a random sample from the exact distribution. One may then use these samples to estimate the predictive distribution using a density estimator such as the kernel estimator.

Gelfand and Smith (1990) show that the first approach to the estimation of the density has a smaller mean squared error and is, therefore, the preferred method. However, the second approach does yield a sample from the predictive distribution which can be used to compute the posterior of a variety of quantities such as the Kullback–Leibler divergence measure (Cook and Weisberg, 1982).

EXAMPLE. *Censored Regression (Continued)*
Having applied the data augmentation algorithm to this data set, we have in the final iteration a sample from $p(\beta_0, \beta_1, \sigma^2 \mid Y)$. Thus, it is straightforward to calculate the predictive distribution for a future observation.

Figure 5.8 presents two curves representing the predictive distributions at temperatures 180° and 130°, respectively, for the motorette data and matching normal curves. (These matching normal curves are not the normal approximation given by the maximum likelihood analysis.) The densities were constructed by drawing histograms of the values generated from the appropriate normal distribution. As can be seen from Figure 5.8, both curves are slightly more skewed than a normal curve, with the curve corresponding to the lower temperature being more skewed. The means of the two distributions are slightly larger than the corresponding quantities computed via maximum likelihood (as reported in Schmee and Hahn, 1979). On

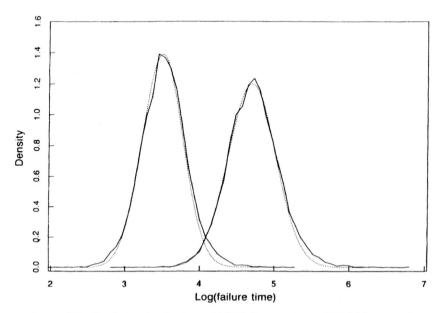

FIGURE 5.8. Predictive distribution at 180° (left curves) and 130° (right curves).

the log scale, the maximum likelihood means are 3.49 and 4.67, respectively, while the data augmentation values are 3.52 and 4.72, respectively. Moreover, the maximum likelihood approach underestimates the variability of the predicted values. For example, the interval obtained by computing the 5th and 95th percentiles of the appropriate predictive distribution at 130° is (on the log scale) given by [4.17, 5.31]. The corresponding normal approximation based interval reported in Schmee and Hahn (1979) is [4.44, 4.90]. This comment reflects the remarks in Aitkin (1981) and Naylor and Smith (1982) that the downward bias that arises in the maximum likelihood estimate of σ produces misleading statements of uncertainty.

5.3 Calculating the Content and Boundary of the HPD Region

As noted in Section 2.6, it is sometimes of interest to delineate a region of the parameter space that contains most of the mass under the posterior. Box and Tiao (1973, p. 122) point out that for a given probability content $1 - \alpha$, the region of highest posterior density (HPD) has smallest volume in parameter space. In addition, these authors point out that a point θ_0 is covered by the highest posterior density region of content $1 - \alpha$ if and only if $P_\theta[p(\theta \mid Y) \geq p(\theta_0 \mid Y) \mid Y] \leq 1 - \alpha$, where the density $p(\theta \mid Y)$ is treated as a random variable. In practice, the content may be computed through the posterior distribution, by some monotonic function of the posterior or by relying on a χ^2 approximation to $-2\ln[p(\theta \mid Y)/p(\hat{\theta} \mid Y)]$, where $\hat{\theta}$ is the posterior mode.

5.3.1. Calculating the Content

The computation of $P_\theta[p(\theta \mid Y) \geq p(\theta_0 \mid Y) \mid Y]$ may not be easy, especially for a high-dimensional parameter space. However, in the context of data augmentation, this probability is easily estimated. Given a sample of size m from $p(\theta \mid Y)$, estimate the probability as

$$\frac{\text{Number of samples such that } p(\theta \mid Y) \geq p(\theta_0 \mid Y)}{m},$$

where θ_0 is a fixed value and $p(\theta \mid Y)$ is evaluated over the m values of θ obtained from the final iteration of data augmentation.

5.3.2. Calculating the Boundary

To compute the boundary of the HPD Region of content $1 - \alpha$ (see Figure 5.9):

1. Locate the 100α percentile of the $p(\theta \mid Y)$ values evaluated at the θ values obtained from the final iteration of the algorithm.
2. Label this value as $p^*(\theta \mid Y)$.

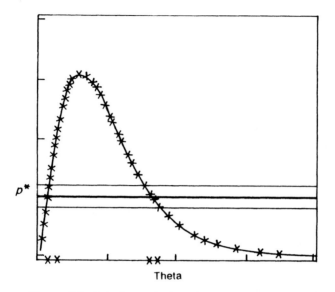

Theta

FIGURE 5.9. Calculating the boundary of an HPD region.

3. Locate those θ values such that the value of the posterior density evaluated at these points is contained in the interval

$$\{p^*(\theta \mid Y) - \varepsilon, \, p^*(\theta \mid Y) + \varepsilon\}.$$

4. Label this set as S_ε.

If $p(\theta \mid Y)$ is unimodal and θ is one dimensional, then for a sufficiently large value of ε there are two distinct groups of values in S_ε. The left-hand endpoint of the HPD in such a case is given by the θ value in the left-hand group of values in S_ε whose posterior is closest to $p^*(\theta \mid Y)$. The right-hand endpoint is the corresponding θ value in the right-hand group.

In higher dimensions, the boundary of the HPD region of content $1 - \alpha$ is approximated by drawing a scatter plot of points in S_ε. One may wish to start with a value of ε large enough to include 3–4% of the θ values and then construct scatter plots for decreasing values of ε. A plot based on a small value of ε will have a smaller bias but will be sparser than a plot based on a larger value of ε. Such a plot may be enhanced by including a smooth curve or surface through the points.

Suppose that θ is partitioned as (θ_1, θ_2), where θ_1 is $k \times 1$ and θ_2 is $(d - k) \times 1$ and interest focuses on θ_1. To compute the content of the HPD region, one must be able to evaluate $p(\theta_1 \mid Y)$. In some situations, one may be able to compute $p(\theta_1 \mid Y)$ by analytic integration of the posterior. Alternatively, it is possible that

$$\frac{1}{m} \sum_{i=1}^{m} p(\theta_1, \theta_2 \mid Y, z_i)$$

may be integrated analytically, while the integration of $p(\theta \mid Y)$ may not be straightforward. This is illustrated in the following example. See also Wei and Tanner (1990b).

EXAMPLE. *Censored Regression (Continued)*
The data augmentation algorithm was implemented with $m = 10,000$ and allowed to run for 15 iterations. Given the output from the final iteration, the computation of the content of the smallest HPD region that includes the value $\sigma^2 = 0.16$, as well as the computation of the boundary of 95% HPD region for σ^2, is straightforward. Note that $p(\sigma^2 \mid Y)$ is readily available since $p(\sigma^2 \mid Y) = \int p(\sigma^2, \beta \mid Y)d\beta$ may be approximated by

$$\frac{1}{m} \sum_{i=1}^{m} \int p(\sigma^2, \beta \mid \text{augmented data set } i)d\beta$$

$$\propto \frac{1}{m} \sum_{i=1}^{m} \frac{s_i^{38}}{(\sigma^2)^{20}} \exp\left(-\frac{19s_i^2}{\sigma^2}\right),$$

where s_i^2 is the least-squares estimate of σ^2 given the ith augmented data set, due to the inverse chi-square/conditional normal factorization.

Given the output from the final iteration of the data augmentation algorithm, one may evaluate the proportion of samples such that $p(\sigma^2 \mid Y) \geq p(\sigma^2 = 0.16 \mid Y)$. For this example, the content of the smallest HPD region that includes the value $\sigma^2 = 0.16$ is equal to 0.94, with a Monte Carlo standard error of 0.0024. In addition, the HPD region of content 95% for σ^2 is given by the interval (0.032, 0.17). This interval was obtained by first evaluating $p(\sigma^2 \mid Y)$ over the σ^2 values obtained in the final iteration of the data augmentation algorithm, next locating the 5th percentile of these $p(\sigma^2 \mid Y)$ values and, finally, locating the "right" and "left" σ^2 values whose posteriors are closest to the 5th percentile. It is noted that the output from the data augmentation algorithm indicates at what height to slice the posterior, and it also provides points that lie on or near the boundary.

The dashed line in Figure 5.10 presents the distribution of the posterior ratio values, $-2\ln[p(\sigma^2 \mid Y)/p(\hat{\sigma}^2 \mid Y)]$, evaluated over the sample σ^2 values obtained from the final iteration of the data augmentation algorithm. The quantity $\hat{\sigma}^2$ is the marginal posterior maximizer, which in this case equals 0.0659. The solid line in Figure 5.10 is the corresponding χ_1^2 approximation. While the solid and dashed curves appear to be in good agreement, it is noted that the area to the right of $-2\ln[p(\sigma^2 = 0.16 \mid Y)/p(\hat{\sigma}^2 \mid Y)]$ under the data augmentation curve is 0.06, while the corresponding value based on the χ^2 approximation is 0.03.

Similar computations regarding the intercept β_0 and the slope β_1 are facilitated by the observation that $p(\beta_0, \beta_1 \mid Y)$ is estimated as

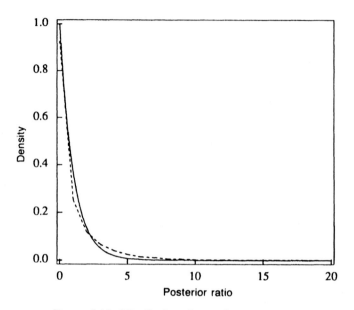

FIGURE 5.10. Distribution of posterior ratio values.

$$\frac{1}{m} \sum_{i=1}^{m} \frac{1}{s_i^2} \left[1 + \frac{(\beta - \hat{\beta}_i)^T X^T X (\beta - \hat{\beta}_i)}{38 s_i^2} \right]^{-20},$$

where $\hat{\beta}_i$ is the least-squares estimate of β given the ith augmented data set and s_i^2 is the corresponding estimate of σ^2. This follows from the inverse chi-square/conditional normal factorization. Figures 5.11(a)–(c) present the 95% HPD region for β_0 and β_1 based on a set S_ε containing 200, 100 and 50 points, respectively. All three plots highlight the high degree of correlation between the two parameters. One may compensate for the sparseness of the last figure by passing a smooth curve through the points.

5.4 Remarks on the General Implementation of the Data Augmentation Algorithm

Two important practical considerations in the implementation of the data augmentation algorithm are the monitoring of convergence and the selection of the number of imputations (m) to be performed per iteration. Tanner and Wong (1987) point out that it is helpful to monitor graphically the progress of the data augmentation

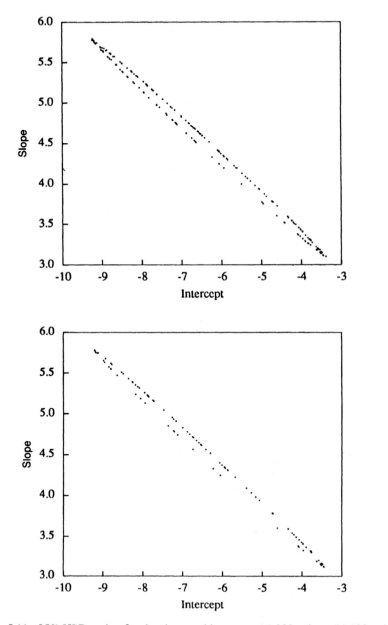

FIGURE 5.11. 95% HPD region for the slope and intercept. (a) 200 points; (b) 100 points.

FIGURE 5.11(c). 95% HPD region for the slope and intercept, 50 points.

algorithm, for example, by using selected percentiles of the estimated posterior distribution. If one is interested in first and second moments, then these moments, rather than extreme tail behavior, may be monitored. The data augmentation algorithm (for a fixed value of m) may be iterated until the fluctuations in such a plot indicate that the process has become stationary. At such a point, the algorithm may be terminated or the value of m increased to improve the precision (with respect to the Monte Carlo variation) of the estimate of the functional of the posterior of interest. In this way, one may start with a smaller value of m and then increase the value of m at various junctures in the iteration process to realize computational savings.

EXAMPLE. *Genetic Linkage (Continued)*
The observed data were taken to be (13, 1, 2, 3). The lines in Figure 5.12 represent the upper, middle and lower quartiles at each iteration. At the initial iteration, m was taken to be 20. The algorithm was then run through 40 iterations, at which point it appeared (see Figure 5.12) that the process had become stationary. The sample size was then increased to 400 and the algorithm proceeded through 20 further iterations. From Figure 5.12, we see that the effect of increasing m has

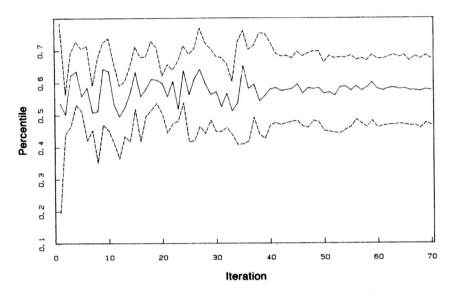

FIGURE 5.12. Upper, middle and lower quartiles versus iteration.

been to reduce substantially the system variability. The final 10 iterations were run with $m = 1600$.

In an alternative approach to monitoring convergence of the algorithm (see Wei and Tanner, 1990c), one monitors the distribution of the weights (w_j) proportional to

$$\frac{p(\theta_j \mid Y)}{g_i(\theta_j)}$$

given the current approximation to the posterior g_i. Clearly, if the current estimate of the posterior is "close" to the posterior, then the distribution of the weights will be degenerate about a constant. In fact, if the weights are normalized to sum to unity, then this constant will be $1/m$. Thus, an alternative graphical approach to monitoring the convergence of the algorithm would be to construct a series of plots, each of which presents the distribution of the weights at a given iteration. A numerical approach would consist of computing some functional of the distribution of the weights, such as the standard deviation, and then to monitor this value as the iterations increase. As the data augmentation algorithm is iterated, for a fixed value of m, the standard deviation will decrease in magnitude. Eventually, the standard deviation will begin to fluctuate about a value. At such a point, one may wish to increase m or to terminate the algorithm depending on the required degree of precision of the estimate of a functional of the posterior. Note that the information on the convergence of *all* the marginals is included in the series of distribution plots of the weights.

EXAMPLE. *Censored Regression (Continued)*
Six histograms are presented in Figure 5.13. The top-left histogram corresponds to the weights (w_j) proportional to

$$\frac{p(\theta_j \mid Y)}{g_2(\theta_j)}$$

obtained after the second iteration, with $m = 100$. The standard deviation of the (normalized) weights $(\Sigma w_j = 1)$ used to construct this plot is 0.017. The remaining histograms correspond to the weights obtained after iterations 5, 7, 10, 12 and 15, all with $m = 100$, with standard deviations of 0.006, 0.006, 0.004, 0.003 and 0.003, respectively. As one can see from the plots, by the 12th iteration, the distribution of the weights (normalized to sum to unity) appears to have degenerated about the value $1/100$. In successive iterations, the distributions of the weights tend to fluctuate about this constant. Having arrived at a stationary process, the parameter values from iterations 13–15 were then pooled. The unweighted average of these 300 β_1 values is 4.40, with a standard error of 0.03. It is noted that if one requires a more precise estimate, then the iterations could be continued with a larger value of m.

5.5 Overview of the Convergence Theory of Data Augmentation

For the examples considered in this chapter, the data augmentation algorithm was seen to converge. In fact, one can state several general results regarding the convergence of the algorithm. See Tanner and Wong (1987) for more details.

Assume that Θ is a connected subspace of $\Re^d$. Let L_1 be the space of Lebesgue integrable functions of θ and $\|f\| = \int | f(\theta) | d\theta$ for $f \in L_1$. The posterior density satisfies

$$g(\theta) = \int K(\theta, \phi)g(\phi)d\phi, \qquad (5.5.1)$$

where $g(\theta) \geq 0$, $\int g(\theta)d\theta = 1$ and $K(\theta, \phi)$ is defined as in Section 5.1. Let $g^*(\theta) = p(\theta \mid Y)$.

The first result shows that the L_1 distances from g^* are nonincreasing in the iterations.

Result 5.5.1. $\|g_{i+1} - g^*\| \leq \|g_i - g^*\|.$

The next result guarantees the uniqueness of the solution to (5.5.1).

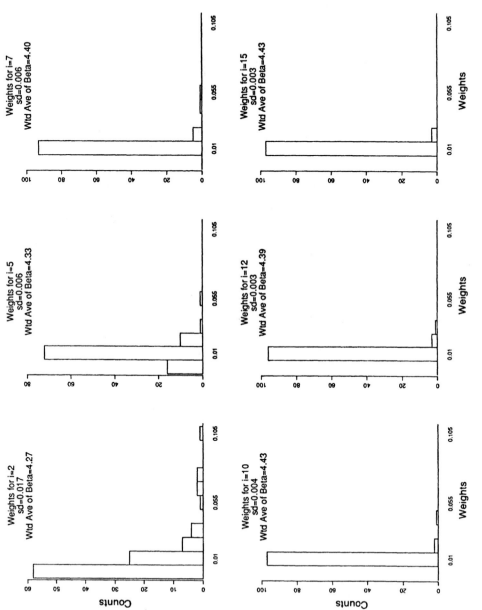

FIGURE 5.13. Distribution of weights across iterations.

Result 5.5.2. Under the regularity condition, g^* is the only density that satisfies (5.5.1).

The final result relates to convergence of $g_{i+1}(\theta) = Tg_i(\theta)$, where $Tf(\theta) = \int K(\theta, \phi) f(\phi) d\phi$.

Result 5.5.3. Under the regularity condition and assuming that the starting value g_0 satisfies $\sup_\theta \mid g_0(\theta)/g^*(\theta) \mid < \infty$, $\|g_i - g^*\| \to 0$.

In Section 5.6, we discuss approaches for obtaining good starting points for data augmentation.

The regularity condition required in Results 5.5.2 and 5.5.3 is

$K(\theta, \phi)$ is uniformly bounded and is equicontinuous in θ. For any $\theta_0 \in \theta$, there is an open neighborhood U of θ_0, so $K(\theta, \phi) > 0$ for all $\theta, \phi \in U$.

The second part of this condition says that if θ and ϕ are close, then it is possible to generate some latent data pattern z from $p(Z \mid \phi, Y)$ such that $p(\theta \mid Z, Y)$ is nonzero. Rosenthal (1993) discusses rates of convergence of data augmentation on finite sample spaces.

5.6 Poor Man's Data Augmentation Algorithms

In this section, we present the poor man's data augmentation algorithms. These algorithms can be used as good starting points for the full data augmentation analysis as well as approximations to the full analysis.

5.6.1. PMDA 1

The data augmentation algorithm states:

a. Generate $z_1, \ldots, z_m \sim p(Z \mid Y)$.
b. Update the approximation to $p(\theta \mid Y)$ as

$$\frac{1}{m} \sum_{i=1}^{m} p(\theta \mid z_i, Y).$$

Note that

$$p(Z \mid Y) = \int_\theta p(Z \mid Y, \theta) p(\theta \mid Y) d\theta = E\{p(Z \mid Y, \theta)\}.$$

Recalling Result 3.2.1 of Section 3.2.1, we have

$$p(Z \mid Y) = p(Z \mid Y, \hat{\theta}) \left\{ 1 + O\left(\frac{1}{n}\right) \right\},$$

where $\hat{\theta}$ is the mode of $p(\theta \mid Y)$. Thus, having obtained $\hat{\theta}$, we have *PMDA 1*.

a. Generate $z_1, \ldots, z_m \sim p(Z \mid Y, \hat{\theta})$.

b. Approximate the posterior as

$$\frac{1}{m} \sum_{i=1}^{m} p(\theta \mid z_i, Y).$$

PMDA 1 is a noniterative algorithm for obtaining an approximation to $p(\theta \mid Y)$. This approximation allows for non-normal shapes for the posterior. It is a "poor man's" algorithm in the sense that it can be used by those who cannot afford the full data augmentation analysis. In addition, it is noted that *PMDA* 1 may provide a good starting point for data augmentation.

There is a connection between *PMDA* and Monte Carlo *EM (MCEM)*. Recall in *MCEM* that at the $(i + 1)$th iteration, the Q function is calculated as

$$\hat{Q}_{i+1}(\theta, \theta^i) = \frac{1}{m} \sum_{j=1}^{m} \log \, p(\theta \mid z_j, Y), \tag{5.6.1}$$

where $z_j \sim p(Z \mid \theta^i, Y)$. When θ^i has converged to the posterior mode $\hat{\theta}$ in *MCEM*, one can drop the logs in (5.6.1) to obtain *PMDA* 1, i.e. an approximation to the entire posterior, not just the maximizer.

EXAMPLE. *Genetic Linkage (Continued)*
To illustrate *PMDA* 1, consider the following small sample data set for the genetic linkage model: (14, 0, 1, 5). The *MCEM* algorithm was run with $m = 5000$ for 15 iterations, yielding $\hat{\theta} = 0.9034$. Using this value of $\hat{\theta}$, 5000 samples were drawn from the conditional predictive distribution $p(Z \mid Y, \hat{\theta})$. Figure 5.14 presents the mixture of augmented posteriors, mixed over the *PMDA* 1 imputations (dotted line), along with the exact observed posterior (solid line). As can be seen from Figure 5.14, *PMDA* 1 successfully recovers the highly skewed shape of the observed posterior.

5.6.2. PMDA-Exact

PMDA 1 is an approximation because $z_i \sim p(Z \mid \hat{\theta}, Y)$ rather than from $p(Z \mid Y)$. If it is straightforward to calculate $p(Z \mid Y)$, then one can use importance sampling to sample from the exact predictive distribution $p(Z \mid Y)$.

PMDA-Exact is given as

a1. Generate $z_1, \ldots, z_m \sim p(Z \mid \hat{\theta}, Y)$.

a2. Calculate

$$w_j = \frac{p(z_j \mid Y)}{p(z_j \mid Y, \hat{\theta})} \, ; j = 1, \ldots, m.$$

b. Calculate the posterior as

$$\sum_{j=1}^{m} w_j p(\theta \mid z_j, Y) \Big/ \sum_{j=1}^{m} w_j.$$

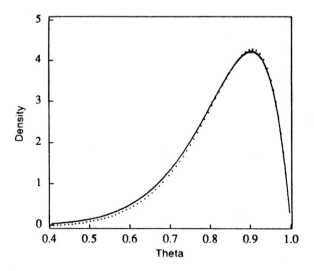

FIGURE 5.14. Marginal of θ. Solid—exact observed posterior, dotted—first-order approximation.

5.6.3 PMDA 2

When $p(Z \mid Y)$ is difficult to compute, a second-order approximation is available. Again, note that

$$p(Z \mid Y) = E[p(Z \mid Y, \theta)] = \int_{\Theta} p(Z \mid Y, \theta) p(\theta \mid Y) d\theta$$

$$= \int_{\Theta} p(Z \mid Y, \theta) \left[\frac{p(\theta \mid Y, Z) p(Z \mid Y)}{p(Z \mid Y, \theta)} \right] d\theta.$$

Recall Result 3.2.2 of Section 3.2.1; namely, via Laplace's method, we have

$$E[g(\theta)] = \left(\frac{\det \Sigma^*}{\det \Sigma} \right)^{1/2} \frac{\exp\{-nh^*(\theta^*)\}}{\exp\{-nh(\hat{\theta})\}} \left[1 + O\left(\frac{1}{n^2} \right) \right],$$

where

$$-nh(\theta) = \log p(\theta \mid Y), \qquad -nh^*(\theta) = -nh(\theta) + \log[g(\theta)],$$

$$\Sigma^* = \left[\frac{\partial^2 h^*}{\partial \theta^2} \Big|_{\theta^*} \right]^{-1}, \qquad \Sigma = \left[\frac{\partial^2 h}{\partial \theta^2} \Big|_{\hat{\theta}} \right]^{-1},$$

$\hat{\theta}$ is the maximizer of $-nh(\theta)$ and θ^* is the maximizer of $-nh^*(\theta)$.

In the present case,

$$-nh(\theta) = \log\, p(\theta \mid Y, Z) - \log\, p(Z \mid \theta, Y) + \log\, p(Z \mid Y)$$

and

$$-nh^*(\theta) = \log\, p(\theta \mid Y, Z) + \log\, p(Z \mid Y).$$

Note that θ^* is the maximizer of $\log\, p(\theta \mid Y, Z)$. Hence, we have shown that to a second-order approximation,

$$p(Z \mid Y) \propto (\det\, \Sigma^*)^{1/2} \frac{p(\theta^* \mid Y, Z)p(Z \mid \hat{\theta}, Y)}{p(\hat{\theta} \mid Y, Z)}.$$

One can use importance sampling to sample from this second-order approximation. In particular, *PMDA* 2 is given as

a1. Generate $z_1, \ldots, z_m \sim p(Z \mid \hat{\theta}, Y)$.
a2. Calculate

$$
\begin{aligned}
w_j &= \frac{(\det\, \Sigma^*)^{1/2} p(\theta_j^* \mid Y, z_j) p(z_j \mid \hat{\theta}, Y)}{p(\hat{\theta} \mid Y, z_j) p(z_j \mid \hat{\theta}, Y)} \\
&= \frac{(\det\, \Sigma^*)^{1/2} p(\theta_j^* \mid Y, z_j)}{p(\hat{\theta} \mid Y, z_j)}.
\end{aligned}
\tag{5.6.2}
$$

b. Approximate the posterior as

$$\sum_{j=1}^{m} w_j\, p(\theta \mid Y, z_j) \Big/ \sum_{j=1}^{m} w_j.$$

EXAMPLE. *Censored Regression (Continued)*
In this example, $\hat{\theta}$ is the posterior mode, i.e. $(\hat{\beta}_0, \hat{\beta}_1, \hat{\sigma})$, $p(Z \mid \hat{\theta}, Y)$ is a conditional normal distribution and $p(\theta \mid Y, Z)$ is given by the inverse chi-square by conditional bivariate normal factorization.

To illustrate the *PMDAs*, we will focus on $p(\sigma^2 \mid Y)$. Following Wei and Tanner (1990a), we approximate

$$p(\sigma^2 \mid Y) = \int p(\sigma^2, \beta_0, \beta_1 \mid Y)d\beta_0 d\beta_1$$

by

$$\frac{1}{m} \sum_{i=1}^{m} \int p(\sigma^2, \beta_0, \beta_1 \mid Y, z_i)d\beta_0 d\beta_1 \propto \frac{1}{m} \sum_{i=1}^{m} \frac{s_i^{38}}{\sigma^{40}} \exp\left(\frac{-19s_i^2}{\sigma^2}\right),$$

where s_i^2 is the least-squares estimate of σ^2 for the ith augmented data set. Thus, *PMDA* 1 is given as

a. Generate $z_1, \ldots, z_m \sim p(Z \mid Y, \hat{\theta})$.

b. Approximate $p(\sigma^2 \mid Y)$ as

$$\frac{1}{m} \sum_{i=1}^{m} \frac{s_i^{38}}{\sigma^{40}} \exp\left(\frac{-19s_i^2}{\sigma^2}\right).$$

PMDA 2 is given as

a1. Generate $z_1, \ldots, z_m \sim p(Z \mid Y, \hat{\theta})$.
a2. Calculate w_j using (5.6.2) above.
 b. Approximate $p(\sigma^2 \mid Y)$ as

$$\sum_{i=1}^{m} w_i \frac{s_i^{38}}{\sigma^{40}} \exp\left(\frac{-19s_i^2}{\sigma^2}\right) \Big/ \sum_{j=1}^{m} w_j.$$

Figure 5.15 presents *PMDA* 1 (short dash line), *PMDA* 2 (long dash line) and the data augmentation estimate (solid line) of the marginal based on $m = 5000$. Note that all three curves are in general accord regarding the location of the mode. However, regarding the shape of the density, the normal approximation to the marginal does not seem appropriate in this case. (In fact, the normal approximation

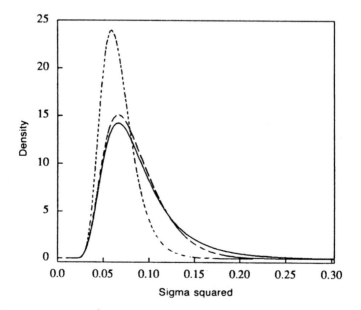

FIGURE 5.15. Marginal of σ^2. Short dash—first-order approximation; long dash—second-order approximation; solid—data augmentation.

is not appropriate even on the $\log(\sigma)$ scale.) *PMDA* 1 gives a hint of the skew in the marginal posterior. *PMDA* 2 does, however, represent an improvement.

In practice, the magnitude of the error of *PMDA* 1 will not be known. Having obtained *PMDA* 1, one may wish to "check" it against *PMDA* 2. If they are in accord, there may be little reason to doubt the validity of either approximation. If, as in this case, they are in discord, it is not clear whether the error in *PMDA* 2 can be neglected. One may then wish to proceed to the implementation of the full data augmentation algorithm using *PMDA* 2 as a starting point (i.e. $g_0(\theta)$) for the data augmentation algorithm.

5.7 Sampling/Importance Resampling (*SIR*)

Rubin (1987a) developed the *SIR* (noniterative) algorithm for drawing missing data patterns from $p(Z \mid Y)$. His intent was to develop methodology to allow for the supplementation of an incomplete public-use database. Standard complete-data methods can then be applied to each of the m augmented data sets to facilitate the analysis of the database. In this context, the fraction of missing information is typically modest. Moreover, one desires a technique that is adequate when m is small. Other imputation methods are presented in Section 5.8.

To begin, one requires a good approximation of $p(\theta, Z \mid Y)$:

$$h(\theta \mid Y)h(Z \mid \theta, Y) = h(\theta, Z \mid Y),$$

where $h(\theta \mid Y)$ approximates $p(\theta \mid Y)$ and $h(Z \mid \theta, Y)$ approximates $p(Z \mid \theta, Y)$.

Step 1. Draw M values of (θ, Z) from $h(\theta, Z \mid Y)$.
Step 2. Calculate

$$w_j = w(\theta_j, z_j \mid Y) \propto \frac{p(Y, z_j \mid \theta_j)p(\theta_j)}{h(\theta_j, z_j \mid Y)} \quad \text{for } j = 1, \ldots, M.$$

This follows since $p(\theta, Z \mid Y) \propto p(\theta)p(Y, Z \mid \theta)$.

Step 3. Draw $m < M$ values of Z from $z_1, \ldots, z_M$ with probabilities proportional to $w_1, \ldots, w_M$.

Cochran (1977, Chapter 9) presents methods for drawing such samples. The adequacy of the approximation can be assessed by examining the distribution of the weights. In particular, a highly skewed distribution indicates a poor approximation.

If one can sample from $p(Z \mid \theta, Y)$; i.e. $h(Z \mid \theta, Y) = p(Z \mid \theta, Y)$, then the weights

$$w_j \propto \frac{p(Y \mid \theta_j)p(z_j \mid \theta_j, Y)p(\theta_j)}{h(\theta_j \mid Y)p(z_j \mid \theta_j, Y)} = \frac{p(Y \mid \theta_j)p(\theta_j)}{h(\theta_j \mid Y)}.$$

Note that in this case the weights do not depend on z_j. Hence, in step 1, only the θ_j need be drawn. In addition, to draw the multiple imputations, draw m values of θ_j with probability proportional to w_j and for each θ_j^* draw z_j^* from $p(Z \mid \theta_j^*, Y)$.

Note that as $M/m \to \infty$, the m pairs (z_j^*, θ_j^*) are selected with probability

$$\frac{h(\theta, Z \mid Y)w(\theta, Z \mid Y)}{\int \int h(\theta, Z \mid Y)w(\theta, Z \mid Y)d\theta dZ}$$

$$= \frac{p(Y, Z \mid \theta)p(\theta)}{\int \int p(Y, Z \mid \theta)p(\theta)d\theta dZ} = p(Z, \theta \mid Y).$$

By "peeling-off" the z^* values, one obtains an *iid* sample of size m from $p(Z \mid Y)$.

The choice of M/m depends on the fraction of missing information. Rubin (1987a) suggests that $M/m = 20$ should be adequate, though he does mention an adaptive scheme for selecting M/m which may warrant further study.

Gelfand and Smith (1990) modify *SIR* to obtain estimates of the functions $p(\theta \mid Y)$ or $p(Z \mid Y)$ rather than to obtain an *iid* sample from $p(Z \mid Y)$. In particular, they replace step 3 with weighted averages.

Step 3.

a. Calculate $p(\theta \mid Y)$ via

$$\sum_{j=1}^{M} w_j p(\theta \mid Y, z_j) \Big/ \sum_{j=1}^{M} w_j.$$

b. Calculate $p(Z \mid Y)$ via

$$\sum_{j=1}^{M} w_j p(Z \mid Y, \theta_j) \Big/ \sum_{j=1}^{M} w_j.$$

By carrying along the weights, $(\theta_1, w_1), \ldots, (\theta_M, w_M)$ is a sample from $p(\theta \mid Y)$, while $(z_1, w_1), \ldots, (z_M, w_M)$ is a sample from $p(Z \mid Y)$.

Gelfand and Smith (1990) examine the performance of the *SIR* algorithm in several examples. They conclude that the performance of the algorithm depends on the specification of $h(\theta, Z \mid Y)$. This conclusion is not unexpected since this dependency is typical of importance sampling techniques.

A similar point based on theoretical considerations is made in the rejoinder of Tanner and Wong (1987). Tanner and Wong (1987) consider the estimation of the posterior moment:

$$\rho = \int a(\theta)p(\theta \mid Y)d\theta.$$

To estimate ρ via data augmentation, one can use

$$\hat{\rho}_{DA} = \frac{1}{M} \sum_{j=1}^{M} a(\theta_j),$$

where $\theta_1, \ldots, \theta_M \sim g_i(\theta)$. To estimate ρ via *SIR*, one can use

$$\hat{\rho}_{SIR} = \sum_{j=1}^{M} w_j a(\theta_j) \Big/ \sum_{j=1}^{M} w_j,$$

where $(\theta_1, z_1), \ldots, (\theta_M, z_M) \sim h(\theta, Z \mid Y)$ and $w_1, \ldots, w_M$ are the weights computed in step 2.

It then follows that

$$E(\hat{\rho}_{DA}) = \int a(\theta) g_i(\theta) d\theta$$

and

$$E(\hat{\rho}_{SIR}) = \int a(\theta) p(\theta \mid Y) d\theta.$$

Hence, $\hat{\rho}_{SIR}$ is unbiased while $\hat{\rho}_{DA}$ is not unbiased, though the bias of $\hat{\rho}_{DA}$ decreases with increasing i since g_i converges to $p(\theta \mid Y)$ in L_1.

Moreover,

$$\text{var}(\hat{\rho}_{DA}) = \frac{1}{M} \left\{ \int a^2(\theta) g_i(\theta) d\theta - (E\hat{\rho}_{DA})^2 \right\}$$

and

$$\text{var}(\hat{\rho}_{SIR}) = \frac{1}{M} \left\{ \int \int a^2(\theta) \frac{p^2(\theta, Z \mid Y)}{h(\theta, Z \mid Y)} dZ d\theta - \rho^2 \right\}.$$

Thus, both variances decrease with increasing M. However, for any fixed M, if $h(\theta, Z \mid Y)$ is a poor approximation to $p(\theta, Z \mid Y)$, $\text{var}(\hat{\rho}_{SIR})$ can be quite large (due to the p^2/h ratio), leading to a low Monte Carlo efficiency of SIR.

What should be done when the weights are highly skewed? In this regard, Tanner and Wong (1987) suggest that the data augmentation algorithm be used to refine the importance function of SIR so that the weights are satisfactorily distributed (hence insuring an adequate performance, i.e. low Monte Carlo standard error of the SIR algorithm). Alternatively, when $h(Z \mid \theta, Y) = p(Z \mid \theta, Y)$ one can potentially use PMDA 1 or 2 as importance functions for the SIR algorithm. These ideas have not been investigated. In summary, we see that iteration corrects for inaccuracies in the starting distribution.

5.8 General Imputation Methods

5.8.1. Introduction

Imputation is a popular approach for handling nonresponse in a survey. In imputation, each missing item is "filled-in" and the survey data are then analyzed using standard techniques for complete data. Rubin (1987b) points out that imputation is advantageous in situations where the data collector and data analyst are different people and the collector has access to more information (due to confidentiality constraints) than the analyst—see also Meng (1994a). The major drawback to the *single* imputation approach is that the imputed values are treated as if known, thus ignoring the variability due to imputation. Rubin (1987b) suggests *multiple* imputation as a device for handling nonresponse. An important contribution of this device is that it allows for the assessment of variation within and between imputations.

In this section, we assume that the scalar responses follow the model $Y_i = X_i^T \theta + \varepsilon_i$, where the ε_i's are *iid* random variables with mean 0 and variance σ^2, θ is the coefficient vector of dimension d and X_i is a covariate vector of known constants.

We consider four techniques for imputing a single set of n_0 missing responses $Y_{(1)}, \ldots, Y_{(n_0)}$ given the observed data

$$\begin{pmatrix} Y_1 \\ X_1 \end{pmatrix}, \ldots, \begin{pmatrix} Y_{n_1} \\ X_{n_1} \end{pmatrix} \text{ and } X_{(1)}, \ldots, X_{(n_0)}.$$

Multiple imputations are obtained by repeated (independent) applications of the particular technique. Like *SIR*, these techniques are noniterative. In contrast with *SIR*, these techniques do not require the specification of the distribution of the ε_i's. However, these techniques are restricted to less complicated situations than *SIR*.

5.8.2. Hot Deck Imputation

Assume that the covariates X take on b distinct values and for each covariate there are several responses. Following Schenker and Welsh (1988), reexpress the data in b blocks as

$$\begin{pmatrix} Y_{11} \\ X_1 \end{pmatrix}, \begin{pmatrix} Y_{12} \\ X_1 \end{pmatrix}, \ldots, \begin{pmatrix} Y_{1n_{11}} \\ X_1 \end{pmatrix}, \underbrace{\begin{pmatrix} * \\ X_1 \end{pmatrix}, \ldots, \begin{pmatrix} * \\ X_1 \end{pmatrix}}_{n_{01}}$$

$$\vdots$$

$$\begin{pmatrix} Y_{b1} \\ X_b \end{pmatrix}, \begin{pmatrix} Y_{b2} \\ X_b \end{pmatrix}, \ldots, \begin{pmatrix} Y_{bn_{1b}} \\ X_b \end{pmatrix}, \underbrace{\begin{pmatrix} * \\ X_b \end{pmatrix}, \ldots, \begin{pmatrix} * \\ X_b \end{pmatrix}}_{n_{0b}}$$

where $n_1 = \sum_{j=1}^{b} n_{1j}$ and $n_0 = \sum_{j=1}^{b} n_{0j}$. In hot deck imputation, within each block, one is to sample independently with replacement the observed responses to form the multiple imputations for the missing responses. For each of the m augmented data sets, one computes the least-squares estimate of the parameter vector $(\hat{\theta}_j)$. The hot deck estimate $(\hat{\theta}_{HD})$ is then given as $(1/m) \sum_{j=1}^{m} \hat{\theta}_j$.

Schenker and Welsh (1988) provide regularity conditions under which

$$(\hat{\theta}_{HD} - \theta) \sim N(0, V).$$

These authors suggest that V may be approximated by

$$\hat{W} \hat{W}_1^{-1} \hat{W} + \left(\frac{m+1}{m} \right) \hat{B},$$

where

$$\hat{W} = \frac{1}{m} \sum_{j=1}^{m} s_j^2 (\psi_1 + \psi_0)^{-1}$$

$$\hat{W}_1 = \frac{1}{m} \sum_{j=1}^{m} s_j^2 \psi_1^{-1}, \quad \psi_1 = \sum_{j=1}^{n_1} X_j X_j^T, \quad \psi_0 = \sum_{j=1}^{n_0} X_{(j)} X_{(j)}^T$$

and

$$\hat{B} = \frac{1}{m-1} \sum_{j=1}^{m} (\hat{\theta}_j - \hat{\theta}_{HD})(\hat{\theta}_j - \hat{\theta}_{HD})^T.$$

5.8.3. Simple Residual Imputation

Let $r_i = Y_i - X_i^T \hat{\theta}_1$, $1 \le i \le n_1$, where $\hat{\theta}_1$ is the least-squares estimate of θ based on the n_1 complete X, Y tuples. In this technique, one is to draw a sample of n_0 residuals, $r_{(1)}, \ldots, r_{(n_0)}$, by sampling with replacement from the set $\{r_1, \ldots, r_{n_1}\}$ and then letting $Y_{(i)} = X_{(i)}^T \hat{\theta}_1 + r_{(i)}$, $1 \le i \le n_0$. As with the hot deck method, for each of the m augmented data sets, one computes the least-squares estimate $(\hat{\theta}_j)$. The simple residual imputation estimate $(\hat{\theta}_{SRI})$ is given by $(1/m) \sum_{j=1}^{m} \hat{\theta}_j$.

Schenker and Welsh (1988) give sufficient conditions under which $(\hat{\theta}_{SRI} - \theta) \sim N(0, V)$. The authors suggest that V can be estimated via $\hat{W}_1 + [(m+1)/m]\hat{B}$, where $\hat{W}_1$ and $\hat{B}$ are defined above.

In effect, this method is a semiparametric extension of $PMDA$ 1. In particular, given the least-squares approximation to the posterior mode, $\hat{\theta}_1$, the SRI method draws latent data patterns from $\hat{p}(Z \mid \hat{\theta}_1, Y)$, where the conditional predictive distribution is nonparametrically estimated using the residuals. The estimate $\hat{\theta}_{SRI}$ is the mean of the mixture

$$\frac{1}{m} \sum_{j=1}^{m} N(\hat{\theta}_j, \hat{\Sigma}_j),$$

where $\hat{\theta}_j$ ($\hat{\Sigma}_j$) is the least-squares estimate of θ (covariance matrix) for the jth augmented data set. Thus, in effect, SRI replaces the augmented posterior of $PMDA$ 1 with a large-sample normal approximation. Also note that the variance of the mixture (conditional on the $\hat{\Sigma}_j$'s and $\hat{\theta}_j$'s) is equal to

$$\frac{1}{m} \sum_{j=1}^{m} (\hat{\Sigma}_j + \hat{\theta}_j \hat{\theta}_j^T) - \left(\frac{1}{m} \sum_{j=1}^{m} \hat{\theta}_j \right) \left(\frac{1}{m} \sum_{j=1}^{m} \hat{\theta}_j \right)^T =$$

$$\text{Average of } [\hat{\Sigma}_1, \hat{\Sigma}_2, \ldots, \hat{\Sigma}_m] + \text{Variance of } [\hat{\theta}_1, \hat{\theta}_2, \ldots, \hat{\theta}_m]. \quad (5.8.1)$$

The first term in (5.8.1) reflects the within-imputation variation. The second term reflects the between-imputation variation, which would vanish if there were no missing observations. The variance estimate given in Schenker and Welsh (1988) includes an inflation factor when estimating the between-imputation variance, i.e.

$$\text{Average of } [\hat{\Sigma}_1, \hat{\Sigma}_2, \ldots, \hat{\Sigma}_m] + \left(1 + \frac{1}{m}\right) \text{Variance of } [\hat{\theta}_1, \hat{\theta}_2, \ldots, \hat{\theta}_m].$$

The inflation factor suggests that the reduction in variability obtained by increasing m is affected mainly by the $1/m$-term. Therefore, using five imputations has relative efficiency larger than 5/6, as compared with using an infinite number of imputations.

5.8.4. Normal and Adjusted Normal Imputation

In contrast to the SRI method, the techniques of this section require sampling of parameter values as well as latent data patterns.

In particular, recall that under a normal error model, the augmented posterior factors into the product of an inverse chi-square and a conditional multivariate normal (Box and Tiao, 1973). In this way, draw σ_*^2 from $(n_1 - d)s_1^2/\chi_{n_1-d}^2$ and θ^* from $N(\hat{\theta}_1, \sigma_*^2\psi_1^{-1})$, where s_1^2 is the mean square error based on the n_1 complete X, Y tuples, and $\hat{\theta}_1$ and ψ_1 are defined above. Next draw n_0 independent observations $e_{(1)}, \ldots, e_{(n_0)}$ from the standard normal distribution and let $Y_{(i)} = X_{(i)}^T\theta^* + \sigma_*e_{(i)}, 1 \le i \le n_0$. For each of the m augmented data sets, compute the least-squares estimate of the parameter vector $(\hat{\theta}_j)$. The normal imputation estimate $(\hat{\theta}_N)$ is given as $(1/m)\sum_{j=1}^m \hat{\theta}_j$. The adjusted normal imputation estimate $(\hat{\theta}_{AN})$ is also obtained by averaging the augmented least-squares estimates. The latter approach differs from the former in that $Y_{(i)} = X_{(i)}^T\theta^* + \sigma_*\bar{r}_{(i)}, 1 \le i \le n_0$, where $\bar{r}_{(i)}$ is the standardized version (mean zero, variance 1) of $r_{(i)}$.

Schenker and Welsh (1988) show that under certain conditions both estimators follow

$$(\hat{\theta} - \theta) \sim N(0, V),$$

where the variance is estimated via $\hat{W}_1 + [(m + 1)/m]\hat{B}$.

The normal and adjusted normal estimators are noniterative adaptations of the data augmentation. In particular, parameter values are generated from the current approximation to the posterior (i.e. the inverse chi-square by multivariate normal factorization). In the normal imputation procedure, the latent data are generated from a normal approximation to $p(Z \mid \theta^*, Y)$, while in the adjusted approach the latent data are drawn from $\hat{p}(Z \mid \theta^*, Y)$, where $\hat{p}$ is the nonparametric estimate of $p(Z \mid \theta^*, Y)$ based on the observed residuals.

While Schenker and Welsh (1988) have considered the large sample properties of these four imputation methods, further work regarding their small sample/small m properties is warranted.

5.8.5. Nonignorable Nonresponse

In this section, we consider the situation where nonresponse is related to the values of the outcome variable. Rubin (1987b) calls this situation *nonignorable nonresponse*. Rubin (1987b) considers two approaches for handling nonignorable nonresponse: mixture models and selection models. For each of these approaches, we consider how to proceed in the absence or presence of followup data.

5.8.5.1. Mixture Model—Without Followup Data

Consider a response Y and a $d \times 1$ covariate vector X. Rubin (1987b) considers an approach for creating multiple imputations for the missing response values. His idea is to factor the posterior distribution of the missing values and the parameters of the model (to be specified below) into three components. The first component is the predictive distribution of the latent data given the parameters. The second component is the conditional distribution of the nonresponders' parameters given the responders' parameters. The third component is the posterior of the responders' parameters given the observed data. Having specified each of these distributions, Rubin (1987b) applies the method of composition to sample latent data patterns.

The normal mixture model supposes that for the responders, Y_i follows the normal distribution with mean $X_i^T \beta_1 + \alpha_1$ and variance σ_1^2. For the nonresponders, Y_i follows the normal distribution with mean $X_i^T \beta_0 + \alpha_0$ and variance σ_0^2. For simplicity, Rubin (1987b) takes $\sigma_1 = \sigma_0 = \sigma$. Hence, under a noninformative prior, the posterior distribution $p(\alpha_1, \beta_1, \sigma \mid Y)$ factors as the product of an inverse chi-square distribution and a conditional normal distribution. Note that in large samples this factorization should be approximately correct even if the Y_i's are not normally distributed. The predictive distribution of the latent data, $p(Z_i \mid \alpha_0, \beta_0, \alpha_1, \beta_1, \sigma, Y)$, is $N(\alpha_0 + X_i^T \beta_0, \sigma^2)$. (Rubin (1987b) assumes that X_i is observed for all units in the sample.) What is the middle distribution linking (α_0, β_0) to (α_1, β_1)?

The conditional distribution of β_0 and α_0 given α_1, β_1 and σ is given by the product of two independent distributions. The conditional marginal of β_0 given α_1, β_1 and σ is a normal distribution with mean β_1 and variance $C_\beta^2 \beta_1 \beta_1^T$. The conditional marginal of α_0 is specified through the average Y value at $\bar{X}_1$ for the nonresponders in the population, $\eta_0 = \alpha_0 + \bar{X}_1^T \beta_0$. The conditional marginal of η_0 is normal with mean $\eta_1 = \alpha_1 + \bar{X}_1^T \beta_1$ (the average Y value for the responders in the population) and variance $C_\eta^2 \eta_1^2$.

Note that when $C_\eta = C_\beta = 0$ these specifications imply an ignorable non-response mechanism. The coefficient C_β specifies the similarity of slopes for responders and nonresponders. Rubin (1987b) refers to C_β as the "prior coefficient of variation" for the nonresponders' regression coefficients. Letting β_{1j} and β_{0j} denote the jth components of β_1 and β_0, respectively, the model supposes that with 95% certainty β_{0j} will lie in the interval $\beta_{1j}(1 \pm 1.96C_\beta)$. The coefficient C_β is taken for all the covariates. Moreover, it is assumed that there is no bias, i.e. the distribution of β_0 is centered at β_1.

The coefficient C_η specifies the similarity of the expected value of Y for the responders and nonresponders with covariate means equal to the covariate mean of the responders. In other words, one is 95% certain that the expected response for the nonresponders with average covariate value $\bar{X}_1$ is contained in the interval $\eta_1(1 \pm 1.96C_\eta)$.

In this way, the joint posterior of the missing data Z, η_0, β_0, η_1, β_1 and σ factors as

$$p(Z \mid \eta_0, \beta_0, \sigma, Y)p(\eta_0, \beta_0 \mid \eta_1, \beta_1, \sigma, Y)p(\eta_1, \beta_1, \sigma \mid Y), \qquad (5.8.2)$$

where the components are defined above.

To implement this normal mixture model, Rubin (1987b) applies the method of composition to (5.8.2). In particular, to sample latent data patterns Rubin performs the following three steps:

Step 1. Draw η_1^*, β_1^*, σ_*^2.

a. Draw σ_*^2 from the inverse χ^2 distribution:

$$(n_1 - d)s_1^2/\chi^2_{(n_1-d)},$$

where s_1^2 is the residual mean square for the responders' regression.

b. Draw β_1^* from the conditional normal:

$$N(\hat{\beta}_1, (X_1^T X_1)^{-1}\sigma_*^2),$$

where $\hat{\beta}_1$ is the least-squares estimate of β_1 and X_1 is the associated design matrix omitting the intercept column.

c. Draw η_1^* from $N(\bar{Y}_1, \sigma_*^2/n_1)$.

Step 2. Draw β_0^* and η_0^*.

a. Draw β_0^* from

$$N(\beta_1^*, C_\beta^2\beta_1^*\beta_1^{*T}).$$

b. Draw η_0^* from

$$N(\eta_1^*, C_\eta^2(\eta_1^*)^2).$$

Step 3. Draw Z from

$$N(\eta_0^* + (X_i - \bar{X}_1)^T\beta_0^*, \sigma_*^2).$$

Steps (1)–(3) are repeated m times to form the augmented data sets.

Rubin (1987b) applies this multiple imputation procedure to data from a 1971 survey of high school principals. Rubin found that in this example the 95% confidence interval for response was insensitive to C_β but quite sensitive to C_η. This reflects the fact the Y is not well predicted by X. This example illustrates how one may use the normal mixture model to display the sensitivity of the analysis to a range of plausible assumptions regarding the "missingness" mechanism. In general, however, this approach does not yield consistent point estimates with infinitely large sample sizes (Brown, 1990) for a particular model.

5.8.5.2. Mixture Model—With Followup Data

In the presence of followup data, one can use this information to generate the latent data patterns—see Glynn, Laird and Rubin (1986, 1993). In particular, for the followups we assume

$$Y = \alpha_0 + X_f^T \beta_0 + \sigma \varepsilon,$$

where n_f is the number of followups, $\varepsilon \sim N(0, 1)$ and X_f is the design matrix for the followups. We have the factorization

$$p(Z, \beta_0, \alpha_0, \sigma \mid Y) = p(Z \mid \beta_0, \alpha_0, \sigma, Y) p(\sigma, \beta_0, \alpha_0 \mid Y). \qquad (5.8.3)$$

Under the noninformative prior, $p(\beta_0, \alpha_0, \sigma \mid Y)$ factors as $s_f^2 (n_f - d) / \chi^2_{(n_f - d)}$ times the conditional normal $N(\hat{\beta}_0, \sigma^2 (X_f^T X_f)^{-1})$, where $\hat{\beta}_0$ is the least-squares estimate using the followup data and s_f^2 is the corresponding residual mean square. Note that in large samples this factorization should be approximately correct even if the Y_i's are not normally distributed. This approach does assume that the followup data are a random sample from the nonresponders. The predictive distribution is $N(\alpha_0 + X_{nf}^T \beta_0, \sigma^2)$, where X_{nf} is the design matrix for the nonfollowups. The predictive distribution can alternatively be approximated using a hot deck type approach on the followup data. This approach would eliminate the need to specify a normal distribution on the nonfollowup responses.

An application of the method of composition to (5.8.3) yields a latent data pattern:

Step 1. Draw α_0^*, β_0^* and σ_*^2 from the inverse chi-square/conditional normal.
Step 2. Draw the latent data for the nonfollowups from $N(\alpha_0^* + X_{nf}^T \beta_0^*, \sigma_*^2)$.

5.8.5.3. Selection Model—Without Followup Data

Greenlees, Reece and Zieschang (1982) introduced the selection model for handling nonignorable nonresponse. Greenlees, Reece and Zieschang (1982) assume that the responses follow

$$Y_i = X_i^T \beta + \sigma \varepsilon_i,$$

where $\varepsilon_i \overset{iid}{\sim} N(0, 1)$. Also assumed in the selection model is that the probability of response depends on Y and other variables V_i: $p(R_i = 1 \mid Y_i, V_i) = 1/[1 + \exp(-\alpha - \gamma Y_i - V_i^T \delta)]$, where $R_i = 1$ if individual i is a responder and 0 otherwise; V_i is a p-dimensional vector of characteristics of individual i; α and γ are scalar parameters; and δ is a $p \times 1$ parameter vector.

For the responders, the contribution to the likelihood is

$$L_i = \frac{1}{1 + \exp(-\alpha - \gamma Y_i - V_i^T \delta)} \frac{1}{\sigma} \phi \left(\frac{Y_i - X_i^T \beta}{\sigma} \right),$$

where $\phi(x)$ is the standard normal density. For the nonresponders, the contribution to the likelihood is

$$L_i = \int_{-\infty}^{\infty} \left\{ 1 - \frac{1}{1 + \exp(-\alpha - \gamma Y - V_i^T \delta)} \right\} \frac{1}{\sigma} \phi \left(\frac{Y - X_i^T \beta}{\sigma} \right) dY.$$

The likelihood of the entire sample is given by the product of the L_i's. Greenlees, Reece and Zieschang (1982) suggest the use of the Gauss–Newton algorithm for maximizing the loglikelihood with respect to α, γ, β, δ and σ.

Having obtained the maximum likelihood estimates $\hat{\alpha}$, $\hat{\beta}$, $\hat{\gamma}$, $\hat{\delta}$ and $\hat{\sigma}$, a single imputation is given as

$$E(Y_i \mid X_i, V_i, R_i = 0) =$$

$$\frac{\int_{-\infty}^{\infty} Y \left\{ 1 - \frac{1}{1 + \exp(-\hat{\alpha} - \hat{\gamma} Y - V_i^T \hat{\delta})} \right\} \frac{1}{\hat{\sigma}} \phi \left(\frac{Y - X_i^T \hat{\beta}}{\hat{\sigma}} \right) dY}{\int_{-\infty}^{\infty} \left\{ 1 - \frac{1}{1 + \exp(-\hat{\alpha} - \hat{\gamma} Y - V_i^T \hat{\delta})} \right\} \frac{1}{\hat{\sigma}} \phi \left(\frac{Y - X_i^T \hat{\beta}}{\hat{\sigma}} \right) dY}.$$

Greenless, Reece and Zieschang (1982) propose the following rejection/acceptance algorithm for drawing multiple imputations from the conditional predictive distribution of the latent data given $\hat{\alpha}$, $\hat{\gamma}$, $\hat{\delta}$, $\hat{\beta}$ and $\hat{\sigma}$.

Step 1. Draw e_i from $N(0, 1)$.
Step 2.

a. Calculate $Y_i = X_i^T \hat{\beta} + \hat{\sigma} e_i$.
b. $p(R = 0 \mid Y_i, V_i) = 1 - 1/\{1 + \exp(-\hat{\alpha} - \hat{\gamma} Y_i - V_i^T \hat{\delta})\}$.

Step 3. Draw u from uniform $[0, 1]$.
Step 4. Accept Y_i if $p(R = 0 \mid Y_i, V_i) \geq u$, otherwise reject and return to step 1.

Greenlees, Reece and Zieschang (1982) apply the selection model to income data from the Current Population Survey. In this case, information from a secondary source allows for a direct test of the method. These authors found that a nonignorable nonresponse model gave more accurate estimates (with regard to mean squared error for the imputed values) than the ignorable nonresponse model.

Rubin (1987b) criticizes the idea of sampling from $p(Z \mid \hat{\alpha}, \hat{\gamma}, \hat{\beta}, \hat{\delta}, \hat{\sigma}, Y)$ rather than from the product

$$p(Z \mid \alpha, \gamma, \beta, \delta, \sigma, Y) p(\alpha, \gamma, \beta, \delta, \sigma \mid Y).$$

Rubin (1987b) shows empirically that the conditional multiple imputation approach underestimates the variability, as would be expected in a first-order-approximation method. Rubin (1987b) suggests approximating $p(\alpha, \gamma, \beta, \delta, \sigma \mid Y)$ by the matching multivariate normal and then applies the method of composition to sample the latent data patterns. Alternatively, one could use the matching multivariate normal as an importance function and then attach the appropriate weights to the latent data patterns. This idea has not been investigated.

5.8.5.4. Selection Model—With Followup Data

The selection model with followup data is quite similar to the model without followup data: If individual i is a responder or nonfollowup nonresponder, then the contribution to the likelihood is as in the previous subsection. For each followup nonresponder, the contribution is

$$\left[1 - \frac{1}{1 + \exp(-\alpha - \gamma Y_i - V_i^T \delta)}\right] \frac{1}{\sigma} \phi \left(\frac{Y_i - X_i^T \beta}{\sigma}\right).$$

5.9 Further Importance Sampling Ideas

5.9.1. Sampling from the Predictive Identity

In step (a1) of the data augmentation algorithm, we draw a θ value from the mixture

$$\frac{1}{m} \sum_{j=1}^{m} p(\theta \mid z_j, Y).$$

Explicitly, we first select a $p(\theta \mid z_i, Y)$ uniformly from the m versions and then sample from this selected augmented posterior $p(\theta \mid z_i, Y)$. In some situations, we may not be able to *directly* sample from the distribution $p(\theta \mid z, Y)$. As such, we need a vehicle to circumvent this problem.

It is straightforward to incorporate importance sampling into the data augmentation algorithm. Given an importance function $I(\theta)$, the predictive identity can be rewritten as

$$p(Z \mid Y) = \int_\Theta p(Z \mid \theta, Y) \frac{p(\theta \mid Y)}{I(\theta)} I(\theta) d\theta.$$

This suggests that to incorporate importance sampling into data augmentation, replace steps (a1) and (a2) of the algorithm with

a1'. Generate ϕ from $I(\phi)$.
a2'. Generate z from $p(Z \mid \phi, Y)$, where ϕ is the value generated in (a1').
a3'. Compute the weight $w = g_i(\phi)/I(\phi)$.

Steps (a1')–(a3') are repeated m times. The current approximation to $p(\theta \mid Y)$ is then updated as the weighted mixture of augmented posteriors of θ, given the augmented data from (a2'), i.e.

$$g_{i+1}(\theta) = \sum_{j=1}^{m} w_j p(\theta \mid z_j, Y) \bigg/ \sum_{j=1}^{m} w_j.$$

Wei and Tanner (1990c) illustrate this idea in the context of censored regression data, where it is difficult to directly sample from $p(\theta \mid Y, Z)$. Their idea is to approximate $p(\theta \mid z_j, Y)$ by $I_j(\theta)$ and then to sample ϕ from $\sum_{j=1}^{m} w_j I_j(\phi) / \sum_{j=1}^{m} w_j = I(\phi)$, which approximates $g_i(\phi)$.

5.9.2. Sequential Imputation

Kong, Liu and Wong (1994) introduce the method of sequential imputation as an alternative means of simulating latent data patterns $\underset{\sim}{z}(1), \ldots, \underset{\sim}{z}(m)$. Their idea is related to the *SIR*, but their algorithm tends to produce more stable importance weights. In many situations, their procedure works well without the need of iteration. Let each latent data pattern $\underset{\sim}{z}(j)$ have n components: $z_1(j), \ldots, z_n(j)$. To draw from $p(z \mid Y)$, sequential imputation imputes the z_i's sequentially and then uses importance sampling weights to avoid iteration. In general, the algorithm starts by drawing z_1^* from $p(z_1 \mid y_1)$ and computing $w_1 = p(y_1)$, where y_1 is the first component of Y. For $t = 2, \ldots, n$, the next two steps are performed sequentially:

a1. Draw $z_t^* \sim p(z_t \mid y_1, z_1^*, \ldots, y_{t-1}, z_{t-1}^*, y_t)$
a2. Compute $p(y_t \mid y_1, z_1^*, \ldots, y_{t-1}, z_{t-1}^*) = p_t$, and let $w_t = w_{t-1} p_t$.

Note that the z_t^*'s are drawn sequentially since each z_t^* is drawn conditioned on the previously imputed missing parts $z_1^*, \ldots, z_{t-1}^*$. Steps (a1) and (a2) are repeated m times to obtain $\underset{\sim}{z}^*(1), \ldots, \underset{\sim}{z}^*(m)$ and $w(1), \ldots, w(m)$ where $\underset{\sim}{z}^*(j) = (z_1^*(j), \ldots, z_n^*(j))$ for $j = 1, \ldots, m$ and

$$w(j) = p(y_1) \prod_{t=2}^{n} p(y_t \mid y_1, z_1^*(j), \ldots, y_{t-1}, z_{t-1}^*(j)).$$

Kong, Liu and Wong (1994) then estimate $p(\theta \mid Y)$ by the weighted mixture

$$\sum_{j=1}^{m} w(j) p(\theta \mid Y, z^*(j)) \Big/ \sum_{j=1}^{m} w(j).$$

From step (a1) we see that $\underset{\sim}{z}^*(j)$ is drawn from

$$p^*(\underset{\sim}{z}^*(j) \mid Y) = p(z_1^*(j) \mid y_1) \times$$

$$\prod_{t=2}^{n} p(z_t^*(j) \mid y_1, z_1^*(j), y_2, z_2^*(j), \ldots, y_{t-1}, z_{t-1}^*(j), y_t)$$

rather than from $p(Z \mid Y)$—the correct distribution. However, notice that the ratio $p(\underset{\sim}{z}^*(j) \mid Y) / p^*(\underset{\sim}{z}^*(j) \mid Y)$ is

$$\frac{p(Y, \underset{\sim}{z}^*(j))}{p(Y)} \quad \frac{p(y_1)}{p(y_1, z_1^*(j))} \prod_{t=2}^{n} \frac{p(y_1, \ldots, y_t, z_1^*(j), \ldots, z_{t-1}^*(j))}{p(y_1, \ldots, y_t, z_1^*(j), \ldots, z_t^*(j))}$$

$$= \frac{p(Y, \underset{\sim}{z^*}(j))}{p(Y)} \frac{p(y_1)}{p(y_1, \ldots, y_n, z_1^*(j), \ldots, z_n^*(j))}$$

$$\times \prod_{t=2}^{n} \frac{p(y_1, \ldots, y_t, z_1^*(j), \ldots, z_{t-1}^*(j))}{p(y_1, \ldots, y_{t-1}, z_1^*(j), \ldots, z_{t-1}^*(j))}$$

$$= \frac{p(y_1)}{p(Y)} \prod_{t=2}^{n} p(y_t \mid y_1, z_1^*(j), \ldots, y_{t-1}, z_{t-1}^*(j)) = \frac{w(j)}{p(Y)}.$$

Since $p(Y)$ does not vary across imputations, the normalized weights $w(j)/\sum_{j=1}^{m} w(j)$ provide the correct adjustment to realize a sample from $p(z \mid Y)$, having actually sampled from $p^*(z \mid Y)$.

Kong, Liu and Wong (1994) illustrate their sequential imputation algorithm on the bivariate normal example of Section 5.1. They realize a comparable plot of the posterior of the correlation coefficient, *without the need to iterate*. In their paper, they also discuss applications to prior sensitivity analysis, case influence, prediction and model selection. Results relating to the efficiency of sequential imputation are presented. Kong, Liu and Wong (1994) also discuss the case where the posterior has to be updated with the arrival of new information. See also Chen and Liu (1996).

5.9.3. Calculating the Posterior

Implicit in the calculation

$$\frac{1}{m} \sum_{j=1}^{m} p(\theta \mid Y, z_j)$$

is the availability of the normalizing constant for $p(\theta \mid Y, Z)$. Chen (1994) presents an importance sampling idea that avoids this issue. Note that

$$p(\theta^* \mid Y) = \int_{\Theta} \int_{Z} w(\theta \mid Z, Y) \frac{p(\theta^*, Z \mid Y)}{p(\theta, Z \mid Y)} p(\theta, Z \mid Y) dZ d\theta$$

for an arbitrary conditional density $w(\theta \mid Z, Y)$, since by Fubini's theorem the double integral is equal to

$$\int_{Z} p(\theta^*, Z \mid Y) \left\{ \int_{\Theta} w(\theta \mid Z, Y) d\theta \right\} dZ = p(\theta^* \mid Y).$$

Thus, Chen (1994) suggests that we approximate $p(\theta^* \mid Y)$ via

$$\frac{1}{n} \sum_{j=1}^{n} w(\theta_j \mid z_j, Y) \frac{p(\theta^*, z_j \mid Y)}{p(\theta_j, z_j \mid Y)}.$$

Note that the normalizing constant for $p(\theta, Z \mid Y)$ is not necessary since it cancels in the ratio $p(\theta^*, Z \mid Y)/p(\theta, Z \mid Y)$. Chen (1994) presents this method for the multicomponent case (see Chapter 6 for a discussion of the Gibbs sampler) and shows that the optimal choice for $w(\theta \mid Z, Y)$ from a minimum variance approach is $p(\theta \mid Z, Y)$. Hence, it is important to choose $w(\theta \mid Z, Y)$ to be close in shape to $p(\theta \mid Z, Y)$.

5.10 Sampling in the Context of Multinomial Data

5.10.1. Dirichlet Sampling

In the genetic linkage example, the augmented posterior is a beta distribution. Thus, it is straightforward to sample from the augmented posterior. In more complicated situations, this simplicity may be absent. We now discuss the Dirichlet sampling procedure, which can be used to sample approximately from the posterior distribution of parametric models for multinomial data.

In the genetic linkage example, given the augmented data, the distribution of the last four cell probabilities (p_2, p_3, p_4, p_5) is equal in distribution to that of $(v_2/2, v_3/2, v_4/2, v_5/2)$, where (v_2, v_3, v_4, v_5) has the Dirichlet distribution proportional to

$$v_2^{x_2} v_3^{x_3} v_4^{x_4} v_5^{x_5}, \tag{5.10.1}$$

where $\sum_{i=2}^{5} v_i = 1$. Note that it is straightforward to sample from such a Dirichlet distribution. The genetic linkage model, however, specifies that (p_2, p_3, p_4, p_5) must lie on the linear parametric curve

$$C = \left\{ \left(\frac{\theta}{4}, \frac{1}{4} - \frac{\theta}{4}, \frac{1}{4} - \frac{\theta}{4}, \frac{\theta}{4} \right) : \theta \subset [0, 1] \right\}.$$

Thus, to sample from the augmented posterior:

a. Observations are drawn from the Dirichlet distribution in (5.10.1).
b. Find the $\hat{\theta}$ that gives cell probabilities $(\hat{p}_2, \hat{p}_3, \hat{p}_4, \hat{p}_5)$ closest to the observed Dirichlet observation (p_2, p_3, p_4, p_5).
c. Accept those points near the parametric curve C e.g. using

$$\left[\sum_{i=2}^{5} (p_i - \hat{p}_i)^2 \right]^{1/2} < \varepsilon,$$

where ε is small. The approximate posterior distribution for θ is obtained by forming the histogram of the corresponding $\hat{\theta}$ values.

This procedure is justified by the following result of Tanner and Wong (1987):

Result. The distribution induced by $p(\theta \mid Y, Z)$ on the curve C is the same as the conditional distribution induced by the Dirichlet distribution in (5.10.1) on C.

In practice, the value of ε is selected by plotting a sequence of estimated posterior distributions of θ corresponding to a sequence of decreasing ε values. This Dirichlet sampling procedure is appropriate when the cell probabilities are linear in θ or if the posterior distribution is relatively concentrated in comparison with the curvature of the parametric surface. Otherwise, the histogram of $\hat{\theta}$ must be multiplied by an adjustment factor.

EXAMPLE. *Genetic Linkage (Continued)*
Assume the augmented data is given by (3,2,2,3). Ten thousand observations were drawn from the Dirichlet distribution corresponding to this data vector. For each Dirichlet observation, the value of θ that gives the closest $(\hat{p}_2, \hat{p}_3, \hat{p}_4, \hat{p}_5)$ vector was found using least squares. The histogram of the 3000 accepted values and the true augmented posterior is given in Figure 5.16. The distribution of the accepted $\hat{\theta}$ values appears to follow the true augmented posterior.

5.10.2. Latent Class Analysis

We now turn our attention to the second example of Chapter 1, latent class analysis.

The data presented in that example represent the response of 3181 participants in the 1972, 1973 and 1974 General Social Surveys, as presented in Haberman (1979). The participants in these surveys were cross-classified by the year of the survey and their responses to each of three questions regarding abortion. Thus, the cell entry n_{abcd} represents the number of subjects who in year $D = d$ give responses

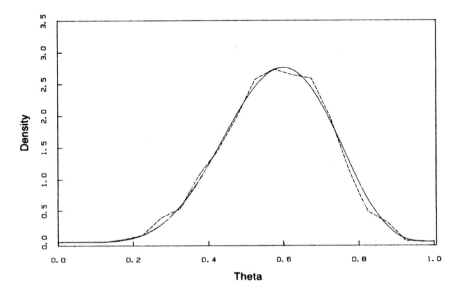

FIGURE 5.16. Marginal of θ.

a to question A, b to question B, and c to question C. Regarding question A, subjects were asked, "Please tell me whether or not you think it should be possible for a pregnant woman to obtain a legal abortion *if she is married and does not want any more children*." In question B, the italicized phrase was replaced with "if the family has a very low income and cannot afford any more children," and in question C it was replaced with "if she is not married and does not want to marry the man." For these data, Haberman (1979) considered several models, one of which is the traditional latent class model.

In this example, the traditional latent class model assumes that the manifest variables (A, B, C, D) are conditionally independent, given a dichotomous latent variable (X). In other words, if the value of the dichotomous latent variable is known for a given participant, then knowledge of the response to a given question provides no further information regarding the responses to either of the other two questions. For the augmented 3×2^4 table, the latent class model specifies that the expected counts m_{xabcd} follow:

$$\log m_{xabcd} = \lambda + \lambda_x^X + \lambda_a^A + \lambda_b^B + \lambda_c^C + \lambda_d^D + \lambda_{xa}^{XA} + \lambda_{xb}^{XB} + \lambda_{xc}^{XC} + \lambda_{xd}^{XD},$$

where

$$\sum_x \lambda_x^X = \sum_a \lambda_a^A = \cdots = \sum_d \lambda_d^D = \sum_x \lambda_{xa}^{XA}$$

$$= \sum_a \lambda_{xa}^{XA} = \cdots = \sum_x \lambda_{xd}^{XD} = \sum_d \lambda_{xd}^{XD} = 0.$$

One parameter of interest associated with this model is the conditional probability of a response a to question A, given that $X = 1$ (which will be denoted as π_{a1}^{AX}). In conjunction with π_{a2}^{AX}, the magnitude of this conditional probability indicates the accuracy of the response a to question A in identifying the latent classification $X = 1$, since the ratio $\pi_{a1}^{AX}/\pi_{a2}^{AX}$ is the likelihood ratio for identifying X based on an observation of A. In the present example, Haberman estimated π_{11}^{AX} to be 0.892.

To obtain the posterior distribution of π_{11}^{AX}, the data augmentation algorithm is implemented as follows. In the initial iteration, the odds of being in the class $X = 2$ (which will be denoted as $\theta_{abcd.}$) is taken to be $\frac{1}{2}$ for all values of a, b, c and d. The unobserved cell counts (n_{abcdx}) are imputed by noticing that, conditional on both $\theta_{abcd.}$ and (the observed cell counts) n_{abcd}, the posterior distribution of n_{abcd1} follows a binomial distribution with parameters n_{abcd} and $1/(1 + \theta_{abcd.})$. The posterior distribution of π_{11}^{AX} is then obtained by drawing from the mixture of augmented posterior distributions. In particular, for a given augmented data set, a vector of probabilities $\{P_{abcdx}\}$ is drawn from the Dirichlet distribution $D(n_{11111}, \dots, n_{22232})$ and some of the observations are discarded using the Euclidean distance criterion, as discussed in Section 5.10.1. The odds of being in the latent class $X = 2$ given that $A = a$, $B = b$, $C = c$ and $D = d$ are updated using the maximum likelihood estimate (under the conditional independence model)

$$\left(\frac{\sum\limits_{b,c,d} p_{abcd2}}{\sum\limits_{b,c,d} p_{abcd1}}\right)\left(\frac{\sum\limits_{a,b,d} p_{abcd2}}{\sum\limits_{a,c,d} p_{abcd1}}\right)\left(\frac{\sum\limits_{a,b,d} p_{abcd2}}{\sum\limits_{a,b,d} p_{abcd1}}\right)\left(\frac{\sum\limits_{a,b,c} p_{abcd2}}{\sum\limits_{a,b,c} p_{abcd1}}\right)\cdot\left(\frac{\sum\limits_{a,b,c,d} p_{abcd1}}{\sum\limits_{a,b,c,d} p_{abcd2}}\right)^3$$

and the algorithm cycles until convergence is achieved. For each augmented data set, the conditional probability of interest is calculated from the equation

$$\pi_{11}^{AX} = \frac{\sum\limits_{b,c,d} p_{1bcd1}}{\sum\limits_{a,b,c,d} p_{abcd1}}.$$

In Figure 5.17, the estimated posterior distribution of π_{11}^{AX} is presented, where the values from the 15th to 20th iterations are pooled ($m = 1600$) to form the histogram in Figure 5.17. As can be seen from Figure 5.17, the posterior distribution appears to be bimodal, with one mode occurring at about 0.039 and the other mode occurring at about 0.886. The reason for this bimodality stems from the unidentifiability inherent in the problem. In the latent class model, the data analyst

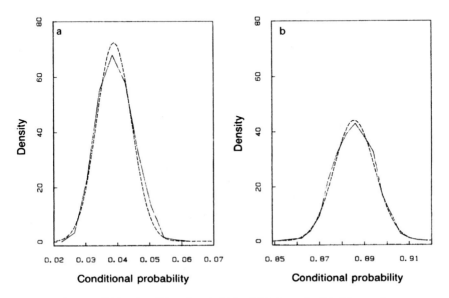

FIGURE 5.17. Conditional probability (a) left mode, (b) right mode.

has the choice of identifying a positive attitude toward abortion with the condition that $X = 1$ or with the condition that $X = 2$. The mode occurring at 0.039 occurs if one identifies a positive attitude with $X = 2$; the second mode occurs if a positive attitude is identified with $X = 1$. In this regard, it is important to note that the modes are well separated. Thus, for the present data set, the conditional probability is, in the Bayesian sense, *locally identifiable*.

Conditioning on the identification of a positive attitude toward abortion with $X = 1$, that is, examining the right mode, our point estimate for π_{11}^{AX} is close to the maximum likelihood estimate (0.886 vs. 0.892). (Such an identification is reasonable given the nature of the question.) In addition, there is little evidence of a departure of the normal approximation from the posterior distribution. Comparing the estimated density to the normal curve with matching mean and standard error (0.009), an overall concordance is observed. A similar conclusion is reached by examining the corresponding rankit plot (see Figure 5.18 (a) and (b)). Regarding the lower mode, some evidence against the normal approximation ($\hat{\mu} = 0.039$, $\hat{\sigma} = 0.006$) is noted. In particular, the posterior distribution is slightly skewed to the right.

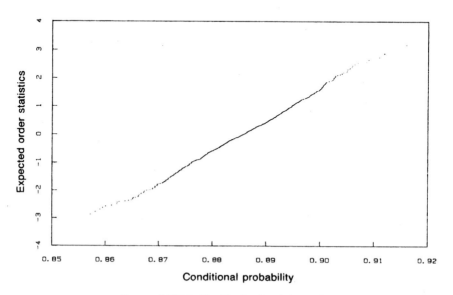

FIGURE 5.18(a). Rankit plot for right mode.

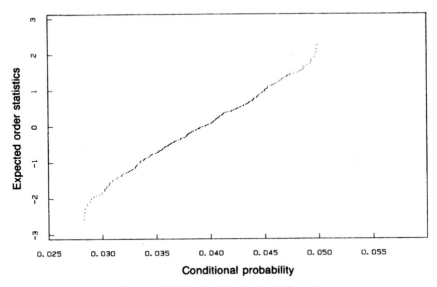

FIGURE 5.18(b). Rankit plot for left mode.

Exercises

1. Recall the genetic linkage model of Section 5.1.
   a. For the data $Y = (125, 18, 20, 34)$ implement the data augmentation algorithm. Use a flat prior on θ. What value did you take for m? How did you assess convergence? (Hint: see Section 5.4.) Compute the posterior mean and posterior variance. Plot the estimated observed posterior along with the normalized likelihood. Use both the simulated values of θ and the mixture of augmented posteriors to estimate the observed posterior.
   b. Repeat (a) for the data $Y = (14, 0, 1, 5)$.
2. For the radiotherapy data of Section 2.3.2, analyze the data using a probit model (via data augmentation). Obtain the posterior marginal for the intercept (α) and for the slope (β). Obtain a 95% HPD region for the slope parameter β. Is $\beta = 0$ in your 95% HPD region? Obtain the posterior distribution of the dose that yields a 50% response rate, i.e. $-\alpha/\beta$. How did you assess convergence? What value did you take for m? Compare your results to the normal approximation results of Section 2.4.
3. a. Repeat Exercise 2 using *PMDA*-1. Compare your results using *PMDA*-1 to those obtained via the full data augmentation approach.
   b. Repeat Exercise 2 using *PMDA*-2. Compare your results using *PMDA*-2 to those using *PMDA*-1 and the full data augmentation approach.

4. For the bivariate normal example of Section 5.1, obtain the posterior distribution of ρ using only the four complete pairs. How does this result compare to the result in Section 5.1?

5. Suppose that X and Y have exponential conditional distributions restricted over the interval $(0, B)$, i.e.

$$p(x \mid y) \propto y e^{-yx}$$

for $0 < x < B < \infty$ and

$$p(y \mid x) \propto x e^{-xy}$$

for $0 < y < B < \infty$, where B is a known constant (see Casella and George, 1992).

 a. Take $m = 1$ and $B = 3$. Run the data augmentation algorithm using these conditionals. (Hint: Reject the exponential deviates that lie outside $(0, B)$.) How did you assess convergence of this chain? Obtain the marginal for x using the mixture of conditionals $p(x \mid y)$, mixed over the simulated y deviates in your chain.

 b. Show that the marginal for x is proportional to $(1 - e^{-Bx})/x$. Compare your result in (a) to this curve.

 c. Repeat (a) and (b) using $B = \infty$. Describe what happens. Is the marginal for x a proper density in this case?

6. a. Use data augmentation to apply the censored regression model of Section 5.1 to the Stanford heart transplant data presented in Chapter 1. Use age as a covariate and take $m = 1$. How did you assess convergence of the algorithm? Obtain the marginal posterior for the regression parameter corresponding to age. Obtain a 95% HPD region for this parameter. Is zero in this range of plausible values?

 b. Evaluate the predictive distribution for the model in (a) at several observed values of the covariate. Compare the means of these distributions to the observed response at these covariate values to assess the adequacy of the model.

 c. Repeat (a) and (b) using the mismatch score as the covariate.

 d. Repeat (a) and (b) using both age and mismatch as covariates.

7. Recall Exercise 8 of Chapter 3. Locate the mode of $p(\alpha, \beta \mid Y)$, i.e. $\hat{\alpha}$ and $\hat{\beta}$. Answer part (f) of that problem by sampling from $p(\theta_{71} \mid \hat{\alpha}, \hat{\beta}, Y)$. Compare your results to those obtained via importance sampling.

6

Markov Chain Monte Carlo: The Gibbs Sampler and the Metropolis Algorithm

6.1 Introduction to the Gibbs Sampler

To motivate the Gibbs sampler, we consider a modification of data augmentation which we will refer to as *chained data augmentation*. The Gibbs sampler turns out to be a multivariate extension of chained data augmentation.

6.1.1. Chained Data Augmentation

To begin, consider the two data augmentation equations:

$$\text{Posterior Equation: } p(\theta \mid Y) = \int_Z p(\theta \mid Y, Z)p(Z \mid Y)dZ$$

$$\text{Predictive Equation: } p(Z \mid Y) = \int_\theta p(Z \mid Y, \theta)p(\theta \mid Y)d\theta.$$

The following algorithm is motivated by applying the method of composition to each of these equations:

a1. Given z^*, generate θ^* from $p(\theta \mid z^*, Y)$.
a2. Generate z^* from $p(Z \mid \theta^*, Y)$ where θ^* is the value obtained in (a1).

The posterior equation suggests that one may draw a value of θ by sampling from $p(\theta \mid z^*, Y)$ where z^* is the value obtained in (a2). Given the new value for θ, step (a2) may be repeated and the algorithm iterated. It is pointed out that this iterative algorithm is data augmentation with $m = 1$. We will refer to this version of data augmentation as *chained data augmentation*. Thus, in chained data augmentation we apply the method of composition to both the predictive and posterior equations to realize an observation from $g_i(\theta)$. In data augmentation, we apply the method of composition m times to the predictive equation to yield $z^{(1)}, \ldots, z^{(m)}$ and then apply the Monte Carlo method to the posterior equation to calculate the current approximation to $p(\theta \mid Y)$.

As noted in Tanner and Wong (1987), the values of θ in chained data augmentation (over the iterations $i = 1, 2, \ldots$) form a Markov chain with transition function equal to $K(\theta, \phi)$, as defined in Section 5.1. Under the regularity conditions of Section 5.5, this is a Markov chain with an equilibrium distribution satisfying the fixed point equation (5.1.2) of Section 5.1. By virtue of the results in Section 5.5, we have

$$\theta^{(i)} \xrightarrow{d} \theta \sim p(\theta \mid Y) \tag{6.1.1}$$

and

$$\frac{1}{t} \sum_{i=1}^{t} f(\theta^{(i)}) \xrightarrow{a.s.} E_p(f(\theta)) \tag{6.1.2}$$

for any integrable function f where $\theta^{(i)}$ is the sampled value at iteration i. Thus, (6.1.1) suggests that if the chain is run to equilibrium, one can use the simulated values as a basis for summarizing $p(\theta \mid Y)$. Since successive $\theta^{(i)}$ are correlated, parallel independent runs of the chain or suitable spacings between realizations will be needed to realize an *iid* sample from $p(\theta \mid Y)$. The second result (6.1.2) suggests that the average of a function of interest over values from a chain yields a consistent estimator of its expectation.

EXAMPLE. *Simulation in Hierarchical Models*
Morris (1987) applied chained data augmentation in the context of hierarchical models. In particular, suppose that d population means $\theta = (\theta_1, \ldots, \theta_d)$ are to be estimated having observed d independent normally distributed sample means $Y = (Y_1, \ldots, Y_d)$, where Y_i given θ_i is *iid* $N(\theta_i, V_i)$, $i = 1, \ldots, d$, and the $V_i = \text{var}(Y_i \mid \theta_i)$ are known. The conjugate prior distribution is taken for each θ_i independently, with A unknown, and θ_i given A are *iid* $N(0, A)$, $i = 1, \ldots, d$. The distribution on the hyperparameter A is $cA^{-1-q/2} \exp(-0.5\lambda/A)$, with known $q > 0$ and $\lambda > 0$.

Morris (1987) notes that these choices lead to proper posterior densities. In particular, the distribution $p(\theta \mid Y, A)$ is normal with the jth component distributed as:

$$N[(1 - B_j)Y_j, V_j(1 - B_j)]$$

$j = 1, \ldots, d$ and $B_j = V_j/(V_j + A)$. The distribution $p(A \mid \theta, Y)$ is a reciprocal chi-squared distribution:

$$\frac{\lambda + \|\theta\|^2}{\chi^2_{d+q}},$$

where $\|\theta\|^2$ denotes the sum of squares. Thus, the chained data augmentation algorithm is given in this case as follows. At iteration i:

1. Sample $\theta_j^{(i)}$ from $N[(1 - B_j^{(i)})Y_j, V_j(1 - B_j^{(i)})]$, with $B_j^{(i)} = V_j/(V_j + A^{(i)})$, $j = 1, \ldots, d$.
2. Sample $A^{(i)}$ from $(\lambda + \|\theta^{(i)}\|^2)/\chi^2_{d+q}$.

Morris (1987) suggests that m chained paths be created to allow for computation of posterior means and posterior distributions. Note that in this case the hyperparameter A plays the role of the "missing" data.

EXAMPLE. *Genetic Linkage (Continued)*
The application of chained data augmentation to this example is quite straightforward. Given z^*, a chain would be constructed by sampling a value for θ (θ^*) from $p(\theta \mid z^*, Y) = \text{Beta}(z^* + X_5 + 1, X_3 + X_4 + 1)$ and then sampling a value for Z (z^*) from $p(Z \mid \theta^*, Y) = \text{Bi}(125, \theta^*/(\theta^* + 2))$. One then iterates between these two steps to obtain a chain.

6.1.2. Multivariate Chained Data Augmentation— The Gibbs Sampler

We now consider a multivariate extension of the chained data augmentation algorithm, which is known as the systematic scan Gibbs sampler. Given the starting point $(\theta_1^{(0)}, \theta_2^{(0)}, \ldots, \theta_d^{(0)})$, this algorithm iterates the following loop:

a. Sample $\theta_1^{(i+1)}$ from $p(\theta_1 \mid \theta_2^{(i)}, \ldots, \theta_d^{(i)}, Y)$
b. Sample $\theta_2^{(i+1)}$ from $p(\theta_2 \mid \theta_1^{(i+1)}, \theta_3^{(i)}, \ldots, \theta_d^{(i)}, Y)$

$$\vdots \qquad \vdots$$

c. Sample $\theta_d^{(i+1)}$ from $p(\theta_d \mid \theta_1^{(i+1)}, \ldots, \theta_{d-1}^{(i+1)}, Y)$.

The vectors $\theta^{(0)}, \theta^{(1)}, \ldots, \theta^{(t)}, \ldots$ are a realization of a Markov chain, with transition probability from θ' to θ,

$$K(\theta', \theta) = p(\theta_1 \mid \theta_2', \ldots, \theta_d', Y) p(\theta_2 \mid \theta_1, \theta_3', \ldots, \theta_d', Y) \times$$
$$p(\theta_3 \mid \theta_1, \theta_2, \theta_4', \ldots, \theta_d', Y) p(\theta_d \mid \theta_1, \ldots, \theta_{d-1}, Y).$$

Chan (1993), Geman and Geman (1984), Liu, Wong and Long (1994, 1995), Roberts and Polson (1994), Schervish and Carlin (1992) and Tierney (1994) present conditions such that:

Result 6.1.1. The joint distribution of $(\theta_1^{(i)}, \ldots, \theta_d^{(i)})$ converges geometrically to $p(\theta_1, \ldots, \theta_d \mid Y)$, as $i \to \infty$.

Result 6.1.2. $\frac{1}{n} \sum_{i=1}^{n} f(\theta^{(i)}) \xrightarrow{a.s.} \int f(\theta) p(\theta \mid Y) d\theta$, as $n \to \infty$.

Result 6.1.3. $\sqrt{n} \left(\frac{1}{n} \sum_{i=1}^{n} f(\theta^i) - \int f(\theta) p(\theta \mid Y) d\theta \right) \xrightarrow{weakly} N(0, \sigma_f^2)$

Amit (1991), Applegate, Kannan and Polson (1990), and Rosenthal (1995) provide bounds for rates of convergence in specific situations. Chen and Schmeiser (1993) and Geyer (1992) discuss the estimation of σ_f^2.

In many applications of the Gibbs sampler, d simulations are performed per iteration corresponding to the individual d components of the parameter $\theta =$

$(\theta_1, \ldots, \theta_d)$. In other cases, the components could be subvectors of θ. In this regard, if *highly correlated components* are treated individually, then convergence can be quite slow. If, however, the correlated components are treated as a block (as in data augmentation), this problem may be mitigated. This issue is treated in detail in Liu (1994) and Liu, Wong and Kong (1994, 1995).

Gelman and Rubin (1992) argue that one should monitor the behavior of several chains starting from points sampled from an overdispersed distribution and that one should check both within- and between-chain variation to assess convergence. (Further discussion of this latter idea is presented in Section 6.3.3.) This overdispersed distribution is constructed in three steps. In step 1, one locates the K modes of $p(\theta \mid Y)$. In step 2, one approximates $p(\theta \mid Y)$ by a weighted mixture of multivariate t distributions—with each t distribution centered at a mode. Gelman and Rubin (1992) suggest that four degrees of freedom on each t should suffice. The inverse information at each mode (possibly multiplied by a scaling factor) determines the overdispersed approximation to $p(\theta \mid Y)$. The weights are selected so that the true posterior matches the approximate posterior at each of the K modes. In step 3, they then use the *SIR* method to obtain the starting points for the chains. This last step makes the starting points "more typical of the target distribution" and should speed up the convergence of the algorithm. Gelman and Rubin (1992) suggest that one draw about 1000 points from the mixture of multivariate t distributions, and then draw 10 importance-weighted resamples—using larger samples when more than one major mode exists. It is noted that the first step assumes that one can locate the (dominant) modes, though this may be difficult in practice. It is also noted that the poor man's data augmentation algorithms may provide a useful approximation to $p(\theta \mid Y)$ from which one can draw starting points for the Gibbs sampler chains.

Geyer (1992) argues that inference can be based on one long run of a Markov chain and presents methods from the time series and operations research literature to quantify the variance of an estimator based on correlated observations. Clearly, several parallel runs or batches within a run can indicate that a single run is not sufficiently long. However, there can never be any guarantee that a run is "long enough". Section 6.3 discusses techniques for monitoring convergence of the Gibbs sampler.

EXAMPLE. *Genetic Linkage (Continued)*
Gelfand and Smith (1990) consider an extension of the genetic linkage example. They present the data as $Y = (14, 1, 1, 1, 5)$ and the model

$$\left[\left(\frac{\theta}{4} + \frac{1}{8} \right), \frac{\theta}{4}, \frac{\eta}{4}, \frac{\eta}{4} + \frac{3}{8}, \frac{1}{2}(1 - \theta - \eta) \right].$$

They consider the augmented data set

$$X = (X_1, \ldots, X_7) \sim \text{Mult} \left[22; \frac{\theta}{4}, \frac{1}{8}, \frac{\theta}{4}, \frac{\eta}{4}, \frac{\eta}{4}, \frac{3}{8}, \frac{1}{2}(1 - \theta - \eta) \right].$$

Gelfand and Smith (1990) let $Z = (X_1, X_5)$. Under the Dirichlet (1, 1, 1) prior for (θ, η), the conditionals are:

$$p(\theta \mid Y, \eta, Z) = (1 - \eta)\text{Beta}(X_1 + Y_2 + 1, Y_5 + 1) \quad \theta \in [0, 1 - \eta]$$
$$p(\eta \mid Y, \theta, Z) = (1 - \theta)\text{Beta}(Y_3 + X_5 + 1, Y_5 + 1) \quad \eta \in [0, 1 - \theta]$$
$$p(Z \mid Y, \theta, \eta) = p(X_1, X_5 \mid Y, \theta, \eta)$$
$$= Bi[Y_1, 2\theta(1 + 2\theta)^{-1}]Bi[Y_4, 2\eta(3 + 2\eta)^{-1}].$$

Using Gaussian quadrature techniques, Gelfand and Smith (1991) find that $E(\theta \mid Y) = 0.520$ and $E(\eta \mid Y) = 0.123$. These authors report the corresponding Tierney–Kadane approximation values of 0.518 and 0.088, respectively. Gelfand and Smith (1991) iterated the Gibbs sampler 10 times and considered 20 replicates of the path. To approximate $E(\eta \mid Y)$, the authors averaged the η component of the 10th iteration over the 20 replicates. This process was then repeated 5000 times. Gelfand and Smith (1991) calculate

$$E(\eta \mid Y) = 0.123$$
$$E(\theta^2 \mid Y) = 0.288$$
$$E(\eta^2 \mid Y) = 0.022$$
$$\text{var}(\theta \mid Y) = 0.018$$
$$\text{var}(\eta \mid Y) = 0.0065$$

These values agree with the exact values based on Gaussian quadrature. Gelfand and Smith (1991) show that by the fourth iteration, the Gibbs sampler correctly computed the area to the left of the 5th, 25th, 50th, 75th and 95th percentage point to two decimal places.

6.1.3. Historical Comments

Geman and Geman (1984) present the Gibbs sampler in the context of spatial processes involving large numbers of variables (e.g. image reconstruction). They consider under which situations the conditional distributions, given "neighborhood" subsets of the variables, uniquely determine the joint distribution. In the situation considered in this chapter, Besag (1974) has shown that if the joint distribution $p(\theta_1, \ldots, \theta_d)$ is positive over its entire domain, then the joint distribution is uniquely determined by the d conditional distributions

$$p(\theta_1 \mid \theta_2, \ldots, \theta_d), \ldots, p(\theta_d \mid \theta_1, \ldots, \theta_{d-1}).$$

The related Metropolis algorithm is discussed in Section 6.5.

Li (1988) applied the Gibbs sampler in the context of multiple imputation. Li suggests that the complete data be partitioned into $d + 1$ parts, $X_0, X_1, \ldots, X_d$, where the observed data are X_0 and $X_1, \ldots, X_d$ is a partition of the missing data. Li assumes that X_i can be sampled from $p(X_i \mid X_j, j \neq i)$. His algorithm is:

Step 1. Sample $X_1^{(0)}, \ldots, X_d^{(0)}$ from some distribution.

Step 2a. Sample $X_1^{(j)}$ from

$$p(X_1 \mid X_0, X_2^{(j-1)}, \ldots, X_d^{(j-1)})$$

b. Sample $X_2^{(j)}$ from

$$p(X_2 \mid X_0, X_1^{(j)} X_3^{(j-1)}, \ldots, X_d^{(j-1)})$$

$$\vdots$$

d. Sample $X_d^{(j)}$ from

$$p(X_d \mid X_0, X_2^{(j)}, \ldots, X_{d-1}^{(j)}).$$

Step (2) is then cycled until the algorithm converges. Li (1988) suggests that multiple paths be considered to check for convergence.

Li (1988) illustrates the method in the context of categorical data, latent variables and censored life data. Li (1988) provides conditions such that the distribution of $(X_1^{(i)}, \ldots, X_d^{(i)})$ converges to $p(X_1, \ldots, X_d)$ geometrically fast. Like Metropolis et al. (1953) and Geman and Geman (1984), Li represents the process as a Markov chain with the joint posterior distribution as the stationary distribution.

Tanner and Wong (1987) present the data augmentation algorithm which is a two component version of the Gibbs sampler. One of the basic contributions of Tanner and Wong (1987) was to develop the framework in which Bayesian computations can be performed in the context of iterative Monte Carlo algorithms. Moreover, in their rejoinder they sketch a Gibbs sampler approach for handling hierarchical models with t errors. Gelfand and Smith (1990) present a review of data augmentation, the Gibbs sampler and the *SIR* algorithm due to Rubin (1987a). These authors apply the approaches to several examples and provide an initial practical comparison of the methods. The papers by Besag et al. (1995), Carlin, Gelfand and Smith (1992), Carlin and Polson (1991), Carlin, Polson and Stoffer (1992), Gelfand et al. (1990), Gelfand and Smith (1991), Gelfand, Smith and Lee (1992), Goldman et al. (1996), Kuo and Yang (1995), Lawrence et al. (1993), and Ramgopal, Laud and Smith (1993), apply the Gibbs sampler to a variety of important statistical problems. This listing is but a small part of the important and interesting published applications of the Gibbs sampler. Volume 55 (No. 1) of the *Journal of the Royal Statistical Society, B*, presents extensive discussion of Markov Chain Monte Carlo methods—see Besag and Green (1993), Frigessi et al. (1993), Gilks et al. (1993) and Smith and Roberts (1993). Spiegelhalter et al. (1995a,b) have developed a general purpose computer program for Bayesian inference using the Gibbs sampler called BUGS. This program is available via anonymous ftp from ftp.mrc-bsu.cam.ac.uk in the directory pub/methodology/bugs. A library of recent MCMC papers is maintained on the World Wide Web at http://www.statslab.cam.ac.uk/~mcmc/html/

6.2 Examples

6.2.1. One-Way Variance Component Model

Consider the 1-way variance component model (see Section 4.3):

$$Y_{ij} = \beta + \mu_i + \varepsilon_{ij},$$

where Y_{ij} is the jth observation from the ith class, β is the grand mean, μ_i is the effect of the ith group and ε_{ij} is the error, where $i = 1, \ldots, a$, $j = 1, \ldots, n_i$ and $\sum_{i=1}^{a} n_i = N$. It is assumed that the ε_{ij} are an *iid* sample from $N(0, \sigma_\varepsilon^2)$ and the μ_i are an independent *iid* sample from $N(0, \sigma_\mu^2)$. Under the prior specification $p(\sigma_\varepsilon^2, \sigma_\mu^2, \beta) = p(\sigma_\varepsilon^2)p(\sigma_\mu^2)p(\beta)$, with

$$p(\sigma_\varepsilon^2) \propto \frac{1}{\sigma_\varepsilon^2}, \; p(\sigma_\mu^2) \propto \frac{1}{\sigma_\mu^2}, \; p(\beta) \propto \; constant,$$

we have (see Gelfand et al. 1990):

a. $p(\beta \mid \mu, \sigma_\varepsilon^2, \sigma_\mu^2, Y) = N(\hat{\beta}, \sigma_\varepsilon^2/N)$
b. $p(\mu_i \mid \beta, \sigma_\varepsilon^2, \sigma_\mu^2, Y) = N(\hat{\mu}_i, c_i\sigma_\varepsilon^2)$
c. $p(\sigma_\varepsilon^2 \mid \beta, \mu, \sigma_\mu^2, Y)$ has the same density as $Ns_\varepsilon^2/\chi_N^2$, where χ_N^2 is a χ^2 random variable on $N \, df$.
d. $p(\sigma_\mu^2 \mid \beta, \mu, \sigma_\varepsilon^2, Y)$ has the same density as as_μ^2/χ_a^2, where χ_a^2 is a χ^2 random variable on $a \, df$,

where $\mu = (\mu_1, \ldots, \mu_a)$, $\hat{\beta} = \dfrac{1}{N} \left(\sum_{i=1}^{a} \sum_{j=1}^{n_i} Y_{ij} - \sum_{i=1}^{a} n_i\mu_i \right)$,

$$\hat{\mu}_i = \frac{n_i}{n_i + \tau} (\bar{y}_{i.} - \beta), \qquad c_i = (n_i + \tau)^{-1},$$

for $i = 1, \ldots a$,

$$s_\varepsilon^2 = \sum_{i=1}^{a} \sum_{j=1}^{n_i} (Y_{ij} - \beta - \mu_i)^2/N, \qquad s_\mu^2 = \sum_{i=1}^{a} \mu_i^2/a,$$

$\tau = \sigma_\varepsilon^2/\sigma_\mu^2$, and $\bar{y}_{i.}$ is the ith sample average.

To illustrate how these full conditionals are obtained, note that $p(\beta, \mu, \sigma_\varepsilon^2, \sigma_\mu^2 \mid Y)$ is proportional to

$$p(\beta, \sigma_\varepsilon^2, \sigma_\mu^2)p(\mu \mid \beta, \sigma_\varepsilon^2, \sigma_\mu^2)p(Y \mid \beta, \sigma_\varepsilon^2, \sigma_\mu^2, \mu).$$

Note that this product is proportional to

$$\frac{1}{(\sigma_\mu^2)^{a/2+1}} \exp\left(\frac{-\mu^T\mu}{2\sigma_\mu^2} \right) \frac{1}{(\sigma_\varepsilon^2)^{N/2+1}} \prod_{i=1}^{a} \prod_{j=1}^{n_i} \exp\left(-\frac{1}{2\sigma_\varepsilon^2} (Y_{ij} - \beta - \mu_i)^2 \right) \quad (*).$$

Suppose that we wish to derive the form for $p(\sigma_\mu^2 \mid \beta, \mu, \sigma_\varepsilon^2, Y)$. Since $p(\sigma_\mu^2 \mid \beta, \mu, \sigma_\varepsilon^2, Y)$ is a function of σ_μ^2, we can ignore all the components in (*) that *do not* depend on σ_μ^2, i.e. everything but

$$\frac{1}{(\sigma_\mu^2)^{a/2+1}} \exp\left(\frac{-\mu^T \mu}{2\sigma_\mu^2}\right)$$

is absorbed into the normalizing constant. Recall that the inverse chi-square density has the form proportional to $\theta^{(-v/2+1)}e^{-1/2\theta}$ (see the Appendix of Chapter 1). Hence, it follows that $p(\sigma_\mu^2 \mid \beta, \mu, \sigma_\varepsilon^2, Y)$ has the same density as as_μ^2/χ_a^2, where χ_a^2 is a χ^2 random variable on a df. It similarly follows that $p(\sigma_\varepsilon^2 \mid \beta, \mu, \sigma_\mu^2, Y)$ has the same density as $Ns_\varepsilon^2/\chi_N^2$.

To derive $p(\beta \mid \mu, \sigma_\varepsilon^2, \sigma_\mu^2, Y)$, note that the only component of (*) that depends on β is

$$\prod_{i=1}^{a}\prod_{j=1}^{n_i} \exp\left(-\frac{1}{2\sigma_\varepsilon^2}(Y_{ij} - \beta - \mu_i)^2\right) = \exp\left(\frac{-\sum_{i=1}^{a}\sum_{j=1}^{n_i}(Y_{ij} - \beta - \mu_i)^2}{2\sigma_\varepsilon^2}\right).$$

If we let $Y_{ij}' = Y_{ij} - \mu_i$, then in vector notation we have

$$\exp\left(\frac{-\sum_{i=1}^{a}\sum_{j=1}^{n_i}(Y_{ij} - \beta - \mu_i)^2}{2\sigma_\varepsilon^2}\right) = \exp\left(\frac{-(Y' - X\beta)^T(Y' - X\beta)}{2\sigma_\varepsilon^2}\right),$$

where X is a column of 1's of length N. Now note that

$$(Y - X\beta)^T(Y - X\beta) = [(Y - X\hat\beta) + (X\hat\beta - X\beta)]^T \times$$
$$[(Y - X\hat\beta) + (X\hat\beta - X\beta)] = (Y - X\hat\beta)^T(Y - X\hat\beta) + (\beta - \hat\beta)^T X^T X(\beta - \hat\beta),$$

where $\hat\beta = (X^TX)^{-1}X^TY$, since $(\beta - \hat\beta)^T X^T(Y - X\hat\beta) = 0 = (Y - X\hat\beta)^T X(\beta - \hat\beta)$. It then follows that $\exp(-(Y' - X\hat\beta)^T(Y' - X\hat\beta)/2\sigma_\varepsilon^2)$ is proportional (as a function of β) to

$$\exp\left(\frac{-(\beta - \hat\beta)^T X^T X(\beta - \hat\beta)}{2\sigma_\varepsilon^2}\right).$$

Since $X^T X = N$, $p(\beta \mid \sigma_\varepsilon^2, \sigma_\mu^2, \mu, Y)$ is normal with mean $\hat\beta$ and variance σ_ε^2/N, where $\hat\beta = (X^TX)^{-1}X^TY' = \frac{1}{N}\sum_{i=1}^{a}\sum_{j=1}^{n_i}(Y_{ij} - \mu_i)$. The full conditional for μ_i is obtained in a similar manner. Gelfand et al. (1990) illustrate this Gibbs sampler using data taken from Box and Tiao (1973)—see Exercise 3 at the end of this chapter.

Hobert and Casella (1993) prove the following result:

Result 6.2.1. Consider the model:

$$Y_{ij} = \beta + \mu_i + \varepsilon_{ij},$$

where $i = 1, \ldots, a$, $j = 1, \ldots, n_i$, the μ_i are an *iid* sample from $N(0, \sigma_\mu^2)$, and the ε_{ij} are an *iid* sample from $N(0, \sigma_\varepsilon^2)$, with independent priors $p(\beta) \propto constant$, $p(\sigma_\varepsilon^2) \propto (\sigma_\varepsilon^2)^{-(b+1)}$ and $p(\sigma_\mu^2) \propto (\sigma_\mu^2)^{-(c+1)}$. The resulting posterior distribution will be proper if and only if $a - 1 > -2c > 0$ and $N + 2c + 2b - 1 > 0$.

(A posterior distribution is proper if $\int L(\theta \mid Y) p(\theta) d\theta < +\infty$.) Thus, the prior $p(\sigma_\mu^2) \propto \sigma_\mu^{-2}$ used above violates the Hobert–Casella condition since $-2c$ is not strictly greater than 0. It then follows that the resulting posterior distribution is not proper. This is reflected in the Markov chain generated by the Gibbs sampler by the fact that $\sigma_\mu^2 = 0$ is an absorbing state. In other words, if the Gibbs sampler detailed above is run "long enough", it will eventually become absorbed by the state $\sigma_\mu^2 = 0$ (noticed by Zeger and Karim, 1991). However, if the chain is not run "long enough" then there will be no evidence of "posterior improperness" in the simulated values, and without the above result we would not be cognizant of a problem. Use of a proper prior would avoid this problem, as does $p(\sigma_\mu^2) = \text{constant}$.

6.2.2. Rat Growth Data

In Chapter 1, data on 30 young rats measured weekly for five weeks were presented. Gelfand et al. (1990) suppose individual straight-line growth curves, i.e.

$$Y_{ij} \sim N(\alpha_i + \beta_i x_{ij}, \sigma^2),$$

where $i = 1, \ldots, 30$; $j = 1, \ldots, n_i = 5$; and x_{ij} denotes the age in days of the ith rat at measurement j. The slopes and intercepts are modelled as

$$\begin{pmatrix} \alpha_i \\ \beta_i \end{pmatrix} \sim N \left[\begin{pmatrix} \alpha_c \\ \beta_c \end{pmatrix}, \Sigma \right]$$

with conditional independence for $i = 1, \ldots, 30$ assumed. Gelfand et al. (1990) take the prior

$$p(\alpha_c, \beta_c; \Sigma^{-1}; \sigma^2) = p(\alpha_c, \beta_c) p(\Sigma^{-1}) \rho(\sigma^2),$$

where

$$p(\alpha_c, \beta_c) = N(\eta, C), \quad p(\Sigma^{-1}) = W[(\rho R)^{-1}, \rho]$$

$$p(\sigma^2) = IG \left(\frac{\upsilon_0}{2}, \frac{\upsilon_0 \tau_0^2}{2} \right),$$

W is the Wishart distribution and IG is the inverse-gamma distribution. Define $\theta_i = (\alpha_i, \beta_i)^T$, X_i is the appropriate design matrix,

$$\bar{\theta} = \frac{1}{30} \sum_{i=1}^{30} \theta_i, \quad D_i^{-1} = \sigma^2 X_i^T X_i + \Sigma^{-1}, \quad V = (30\Sigma^{-1} + C^{-1})^{-1},$$

$$\mu_c = (\alpha_c, \beta_c)^T$$

and

$$n = \sum_{i=1}^{30} n_i = 30 \times 5 = 150.$$

The Gibbs sampler is specified by the following distributions:

$$p(\theta_i \mid Y, \mu_c, \Sigma^{-1}, \sigma^2) = N[D_i(\sigma^{-2} X_i^T Y_i + \Sigma^{-1} \mu_c), D_i]$$

$$p(\mu_c \mid Y, \theta_1, \ldots, \theta_{30}, \Sigma^{-1}, \sigma^2) = N[V(30\Sigma^{-1}\bar{\theta} + C^{-1}\eta), V]$$

$$p(\Sigma^{-1} \mid Y, \theta_1, \ldots, \theta_{30}, \mu_c, \sigma^2)$$

$$= W\left[\left[\sum_i (\theta_i - \mu_c)(\theta_i - \mu_c)^T + \rho R\right]^{-1}, 30 + \rho\right]$$

$$p(\sigma^2 \mid Y, \theta_1, \ldots, \theta_{30}, \mu_c, \Sigma^{-1})$$

$$= IG\left\{\frac{n + v_0}{2}, \frac{1}{2}\left[\sum_i (Y_i - X_i\theta_i)^T (Y_i - X_i\theta_i) + v_0\tau_0^2\right]\right\}.$$

For the rat data, Gelfand et al. (1990) take

$$C^{-1} = 0, v_0 = 0, \rho = 2 \text{ and } R = \begin{pmatrix} 100 & 0 \\ 0 & 0.1 \end{pmatrix}$$

reflecting the vague prior information. Note that to sample from the inverse gamma distribution, one draws a deviate from the corresponding gamma distribution and forms the reciprocal of this deviate.

For these data, Gelfand et al. (1990) report convergence of the Gibbs sampler in about 35 iterations with $m = 50$. Figure 6.1 presents the resulting posterior

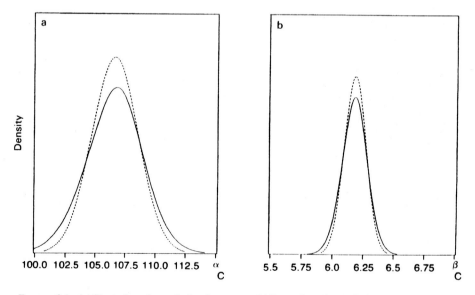

FIGURE 6.1. (a) Posterior of population intercept. (b) Posterior of population slope. Solid— Gibbs; dashed—*EM*. Source: Gelfand et al. (1990).

distribution for α_c and β_c, respectively. The solid line is obtained from the Gibbs sampler, and the dashed line is the normal approximation based on the results of the *EM* algorithm. As can be seen in the plots, the normal approximation underestimates the dispersion of the parameters. An even poorer performance of the normal approximation is noted by Gelfand et al. (1990) for smaller versions of this data set. These lower-dimensional marginals (e.g. $p(\alpha_c \mid Y)$) were obtained by averaging $p(\alpha_c \mid \theta, \beta_c, \Sigma^{-1}, \sigma, Y)$ over the sampled values of θ, μ_c, Σ^{-1} and σ (see Section 5.2). Gelfand and Smith (1990) and Liu, Wong and Kong (1994, 1995) show that this estimate has a smaller mean-square error than nonparametric estimates (e.g. histogram) based on the realized values. Gelfand and Smith (1990) use a Rao–Blackwell argument to derive their result. Hence, this estimate of the marginal is referred to as the *Rao–Blackwell* estimate. The motivation for this estimate is simply applying the method of Monte Carlo to the integral:

$$p(\theta_1 \mid Y) = \int p(\theta_1 \mid \theta_2, \ldots, \theta_d, Y) p(\theta_2, \ldots, \theta_d \mid Y) d\theta_2 \ldots d\theta_d.$$

6.2.3. Poisson Process with a Change Point

Carlin, Gelfand and Smith (1992) consider the application of the Gibbs sampler to a Poisson process model with a change point. In particular, Carlin, Gelfand and Smith (1992) let $Y_i \sim$ Poisson $(\theta t_i), i = 1, \ldots, k$, $Y_i \sim$ Poisson $(\lambda t_i), i = k + 1, \ldots, n$ at the first stage. At the second stage, the authors put independent priors on k, θ and λ: k discrete uniform on $\{1, \ldots, n\}$; $\theta \sim G(a_1, b_1)$ and $\lambda \sim G(a_2, b_2)$, where $G(a, b)$ denotes the gamma density proportional to $\theta^{a-1} e^{-\theta/b}$. At the third stage, the authors let $b_1 \sim IG(c_1, d_1)$ independent of $b_2 \sim IG(c_2, d_2)$, where IG denotes the inverse gamma distribution, and suppose that a_1, a_2, c_1, c_2, d_1 and d_2 are known. The conditional distributions are

$$p(\theta \mid Y, \lambda, b_1, b_2, k) \sim G\left[a_1 + \sum_{i=1}^{k} Y_i, \left(\sum_{i=1}^{k} t_i + b_1^{-1}\right)^{-1}\right]$$

$$p(\lambda \mid Y, \theta, b_1, b_2, k) \sim G\left[a_2 + \sum_{i=k+1}^{n} Y_i, \left(\sum_{i=k+1}^{n} t_i + b_2^{-1}\right)^{-1}\right]$$

$$p(b_1 \mid Y, \theta, \lambda, b_2, k) \sim IG[a_1 + c_1, (\theta + d_1^{-1})^{-1}]$$
$$p(b_2 \mid Y, \theta, \lambda, b_1, k) \sim IG[a_2 + c_2, (\lambda + d_2^{-1})^{-1}]$$

and

$$p(k \mid Y, \theta, \lambda, b_1, b_2) = \frac{L(Y, k, \theta, \lambda)}{\sum_{k=1}^{n} L(Y, k, \theta, \lambda)},$$

where

$$L(Y, k, \theta, \lambda) = \exp\left[(\lambda - \theta) \sum_{i=1}^{k} t_i\right] (\theta/\lambda)^{\sum_{i=1}^{k} Y_i}.$$

TABLE 6.1.   British Coalmining Disaster Data by Year, 1851–1962.

Year	Count	Year	Count	Year	Count	Year	Count
1851	4	1881	2	1911	0	1941	4
1852	5	1882	5	1912	1	1942	2
1853	4	1883	2	1913	1	1943	0
1854	1	1884	2	1914	1	1944	0
1855	0	1885	3	1915	0	1945	0
1856	4	1886	4	1916	1	1946	1
1857	3	1887	2	1917	0	1947	4
1858	4	1888	1	1918	1	1948	0
1859	0	1889	3	1919	0	1949	0
1860	6	1890	2	1920	0	1950	0
1861	3	1891	2	1921	0	1951	1
1862	3	1892	1	1922	2	1952	0
1863	4	1893	1	1923	1	1953	0
1864	0	1894	1	1924	0	1954	0
1865	2	1895	1	1925	0	1955	0
1866	6	1896	3	1926	0	1956	0
1867	3	1897	0	1927	1	1957	1
1868	3	1898	0	1928	1	1958	0
1869	5	1899	1	1929	0	1959	0
1870	4	1900	0	1930	2	1960	1
1871	5	1901	1	1931	3	1961	0
1872	3	1902	1	1932	3	1962	1
1873	1	1903	0	1933	1		
1874	4	1904	0	1934	1		
1875	4	1905	3	1935	2		
1876	1	1906	1	1936	1		
1877	5	1907	0	1937	1		
1878	5	1908	3	1938	1		
1879	3	1909	2	1939	1		
1880	4	1910	2	1940	2		

Carlin, Gelfand and Smith (1992) apply this model to the British coalmining disaster data given in Table 6.1.

For these data, Carlin, Gelfand and Smith (1992) adopt vague higher-order priors by taking $a_1 = a_2 = 0.5$, $c_1 = c_2 = 0$ and $d_1 = d_2 = 1$. Convergence was obtained after 15 iterations with $m = 100$. Figure 6.2 presents $p(k \mid Y)$ for the entire data set (solid line) and for the data set with every fifth year deleted (dashed lines). In both cases, the mode is at $k = 41$. Note that the posterior probability that $k = n$, i.e. there is no change point, is close to 0. Thus, there is strong evidence of a

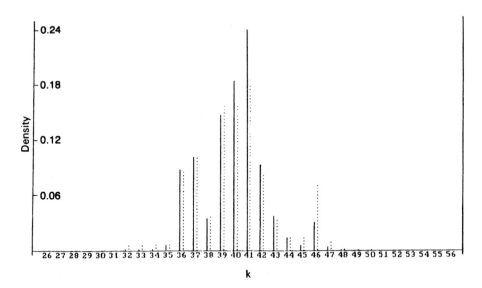

FIGURE 6.2. Posterior distribution of change point. Source: Carlin, Gelfand and Smith (1992).

change. Carlin, Gelfand and Smith (1992) also present the $p(\theta \mid Y)$ and $p(\lambda \mid Y)$ marginals for these data.

6.2.4. Generalized Linear Models with Random Effects

Zeger and Karim (1991) consider the generalized linear model with random effects. In particular, the data consist of the response y_{ij} and a $p \times 1$ vector of predictors x_{ij} for observations $j = 1, 2, \ldots, n_i$ within clusters $i = 1, \ldots, I$. Conditional on a random variable b_i, y_{ij} follows an exponential family distribution

$$f(y_{ij} \mid b_i) = \exp\{[y_{ij}\theta_{ij} - a(\theta_{ij}) + c(y_{ij})]/\phi\}.$$

This is the well-known generalized linear model (McCullagh and Nelder, 1989). The conditional moments $u_{ij} = E(y_{ij} \mid b_i) = \alpha'(\theta_{ij})$ and $v_{ij} = \text{var}(y_{ij} \mid b_i) = a''(\theta_{ij})\phi$ satisfy

$$h(u_{ij}) = \eta_{ij} = x_{ij}^T\beta + z_{ij}^Tb_i$$

and

$$v_{ij} = g(u_{ij})\phi,$$

where h and g are known link and variance functions, respectively; z_{ij} is a $q \times 1$ subset of x_{ij} with random coefficients, β is a vector of regression coefficients and

b_i is a $q \times 1$ vector of random effects following a multivariate normal distribution with mean 0 and variance D.

To complete the specification of the model, let $y_i = (y_{i1}, \ldots, y_{in_i})^T$, $X_i = (x_{i1}, \ldots, x_{in_i})^T$, $Z_i = (z_{i1}, \ldots, z_{in_i})^T$, $u_i = (u_{i1}, \ldots, u_{in_i})^T$, $\eta_i = (\eta_{i1}, \ldots, \eta_{in_i})^T$ and $V_i = \mathrm{diag}(v_{i1}, \ldots, v_{in_i})$.

Zeger and Karim (1991) apply the Gibbs sampler in this context. To implement the Gibbs sampler, these authors consider three distributions: $p(\beta \mid b, D, Y)$; $p(D \mid \beta, b, Y)$ and $p(b \mid \beta, D, Y)$.

Regarding $p(\beta \mid b, D, Y)$, Zeger and Karim (1991) show that $p(\beta \mid b, D, Y) = p(\beta \mid b, Y)$. Given the values $b^{(K)}$ for the b's, the random-effects model reduces to a generalized linear model with offset $z_{ij} b_i^{(K)}$ for each response. For a flat prior, $p(\beta \mid b^{(K)}, Y)$ is proportional to the likelihood $\prod_{i,j} f(y_{ij} \mid b_i^{(K)})$. To sample from the likelihood, Zeger and Karim (1991) employ a rejection/acceptance algorithm. (See Section 3.3.3 for a discussion of this algorithm.)

In the context of $p(\beta \mid b^{(K)}, Y)$, Zeger and Karim (1991) take the majorizing function of the rejection/acceptance algorithm to be $c_1 N(\hat\beta^{(K)}, c_2 V_{\hat\beta}^{(K)})$, where $\hat\beta^{(K)}$ is the maximizer of $p(\beta \mid b^{(K)}, Y)$ and $V_{\hat\beta}^{(K)}$ is the inverse Fisher information. These quantities may be obtained by performing the generalized linear model regression of y_{ij} on x_{ij} with $z_{ij} b_i^{(K)}$ taken as offsets. Zeger and Karim (1991) inflate the variance of the majoring function by taking $c_2 = 2.0$. The quantity c_1 is selected so that the modes of the majorizing density and $p(\beta \mid b^{(K)}, Y)$ are equal.

Regarding the conditional $p(b \mid \beta^{(K)}, D^{(K)}, Y)$, the authors again use a rejection/acceptance algorithm. In this case, the majorizing function is $cN(A, B)$, where $A = (Z_i^T V_i Z_i + D^{-1})^{-1} Z_i^T V_i (y_1^* - X_i \beta)$; $B = c_2 (Z_i^T V_i Z_i + D^{-1})^{-1}$; y_i^* is the linear approximation to $h(y_i)$:

$$y_i^* = \eta_i + \left(\frac{\partial u_i}{\partial \eta_i}\right)^{-1} (y_i - u_i)$$

and

$$\eta_i = X_i \beta + Z_i b_i.$$

Zeger and Karim (1991) show that $p(D \mid \beta, b, Y) = p(D \mid b)$. Assuming that the b_i's are independent normal with mean 0 and variance D and adopting the noninformative prior, the posterior of D^{-1} is a Wishart distribution with parameters

$$S^{(K)} = \sum_{i=1}^{I} b_i^{(K)} b_i^{(K)^T}$$

and $I - q + 1$ degrees of freedom.

Zeger and Karim (1991) applied the algorithm to fit a logistic-normal random-effects model to infectious disease data on 250 Indonesian children. The children were examined up to six times for the presence of respiratory infection. There were a total of 1200 responses. The covariates considered are: age in months (centered at 36); presence/absence of xerophthalmia; cosine and sine terms for the annual cycle; gender; height for age as a percentage of the National Center for Health

Statistics (NCHS) standard (centered at 90%) and presence of stunting, defined as being below 85% in height for age. The intercept is taken as a random effect with a normal distribution.

Table 6.2, taken from Zeger and Karim (1991), lists the estimated posterior mean, mode, standard deviation and 90% confidence interval for each parameter. These statistics were computed from 1000 simulated values obtained after the algorithm converged. As can be seen from Table 6.2, respiratory infection appears to be related to age, gender, season and height for age. Further analysis, including a presentation of the marginal densities, is given in Zeger and Karim (1991). See also Carter and Kohn (1994), Dellaportas and Smith (1993) and Wakefield et al. (1994).

6.2.5. Item Response Model

Having presented the Newton–Raphson solution in Section 2.3.2 and the *EM* solution in Section 4.1, we now present the Gibbs sampler/data augmentation solution to the probit version of the item response model. Recall that n subjects each take an examination with k items. Each item is scored as correct or incorrect and Y_{ij} equals 1 if the ith student correctly answers the jth item and 0 otherwise. Let Y_{ij} be independent Bernoulli random variables where the probability of success is $p_{ij} = \Phi(\alpha_j \theta_i - \beta_j)$, where $\Phi(.)$ is the standard normal cdf. As before, the quantity θ_i represents the unobserved "ability" of the ith student, while α_j measures the discriminatory power (to discriminate between strong and weak students) of the jth item and β_j represents the difficulty of the jth item.

TABLE 6.2. Posterior Characteristics.

Explanatory variable	Mean	Mode	St. dev.	90% C.I.
Intercept	−2.74	−2.72	0.23	−3.10, −2.34
Age (months)	−0.04	−0.04	0.01	−0.05, −0.02
Xerophthalmia (0—no; 1—yes)	0.64	0.64	0.51	−0.31, 1.34
Sex (0—male, 1—female)	−0.61	−0.57	0.18	−0.91, −0.33
Seasonal cosine	−0.17	−0.16	0.17	−0.45, 0.13
Seasonal sine	−0.45	−0.44	0.26	−0.89, −0.03
Height for age (% of NCHS standard)	−0.05	−0.05	0.03	−0.10, −0.01
Stunted (<85% height for age; 0—no, 1—yes)	0.18	0.24	0.47	−0.63, 0.91
D	0.80	0.78	0.40	0.25, 1.54

Albert (1992) introduces a vector of independent (latent) random variables $Z = (Z_{11}, \cdots Z_{nk})$, where Z_{ij} (corresponding to Y_{ij}) is distributed as a normal random variable with mean $\eta_{ij} = \alpha_j \theta_i - \beta_j$ and a variance of unity. He takes $Y_{ij} = 1$ if $Z_{ij} > 0$ and $Y_{ij} = 0$ otherwise, i.e. if the latent variable crosses the threshold, the student answers the question correctly. It then follows that the Y_{ij} are independent Bernoulli random variables with probability of success $p_{ij} = p(Y_{ij} = 1 \mid \theta_i, \alpha_j, \beta_j) = \Phi(\alpha_j \theta_i - \beta_j)$.

Recall the notation $\theta = (\theta_1, \cdots \theta_n)$ and $\zeta = (\zeta_1, \ldots, \zeta_k)$, where $\zeta_j = (\alpha_j, \beta_j)$. To implement the Gibbs sampler for the item response model, we need to derive the three full conditionals: $p(Z \mid \zeta, \theta, Y)$, $p(\zeta \mid \theta, Z, Y)$ and $p(\theta \mid Z, \zeta, Y)$. We first derive $p(Z \mid \zeta, \theta, Y)$. When Y_{ij} equals unity (i.e. the ith student correctly answers the jth question), the conditional distribution of the latent data $p(Z_{ij} \mid \theta, \zeta, Y)$ is normal with mean η_{ij} and variance unity *truncated on the left* by 0 (since the latent variable crossed the threshold). When Y_{ij} equals zero, the conditional predictive distribution $p(Z_{ij} \mid \theta, \zeta, Y)$ is normal with mean η_{ij} and variance unity *truncated on the right* by 0 (since the latent variable did not cross the threshold).

Albert (1992) shows (see Exercise 6 of this chapter) that $p(\theta_i \mid Z, \zeta, Y)$ is normal with mean

$$\frac{\hat{\theta}_i}{1 + V}$$

and variance

$$\frac{V}{1 + V},$$

where

$$\hat{\theta}_i = V \sum_{j=1}^{k} \alpha_j (Z_{ij} + \beta_j) \text{ and } V = \left[\sum_{j=1}^{k} \alpha_j^2 \right]^{-1}.$$

He also shows that $p(\zeta_j \mid \theta, Z, Y)$ is the bivariate normal distribution restricted over the range $\alpha_j > 0$, with mean $\hat{\zeta}_j = (X^T X)^{-1} X^T Z_j$ and variance $(X^T X)^{-1}$, where $X = [\theta, -\mathbf{1}]$, $\mathbf{1}$ is an $n \times 1$ vector of 1's and $Z_j = [Z_{1j}, \ldots, Z_{nj}]^T$. Albert (1992) illustrates this approach using data from a mathematics placement examination.

6.3 Assessing Convergence of the Chain

6.3.1. The Gibbs Stopper

The basic idea behind the Gibbs stopper is to assign the weight w to the vector $\theta = (\theta_1, \ldots, \theta_d)$, which has been drawn from the current approximation to the joint density g_i via

$$w(\theta) = \frac{q(\theta_1, \ldots, \theta_d \mid Y)}{g_i(\theta_1, \ldots, \theta_d)},$$

where $q(\theta_1, \ldots, \theta_d \mid Y)$ is proportional to the posterior density $p(\theta_1, \ldots, \theta_d \mid Y)$. As g_i converges toward $p(\theta_1, \ldots, \theta_d \mid Y)$, the distribution of the weights (associated with values of θ drawn from g_i) should converge toward a spike distribution. This observation has been found useful in assessing convergence of the Gibbs sampler as well as in transforming a sample from g_i into a sample from the exact distribution; see Ritter and Tanner (1992) and Gelfand and Sahu (1994). Historically, the idea of using importance weights to monitor convergence of the data augmentation algorithm was first presented in the rejoinder of Tanner and Wong (1987) and illustrated in Wei and Tanner (1990c), as discussed in Section 5.4.

If $q(\theta \mid Y) = f(Y \mid \theta)p(\theta)$, then as $g_i(\theta)$ converges to

$$p(\theta \mid Y) = \frac{f(Y \mid \theta)p(\theta)}{\int f(Y \mid \theta)p(\theta)d\theta} = \frac{f(Y \mid \theta)p(\theta)}{p(Y)},$$

the weights converge to $p(Y)$. This quantity $p(Y)$ is called the *marginal likelihood* and is used in the computation of the Bayes factor (see Kass and Raftery, 1995). Chib (1995) presents other approaches for computing $p(Y)$.

To write down the functional form for g_i for the Gibbs sampler, we introduce notation following Schervish and Carlin (1992). Let $p^{(i)}(\theta) = p(\theta_i \mid \theta_1, \ldots, \theta_{i-1}, \theta_{i+1}, \ldots, \theta_d, Y)$. For two vectors θ and θ', define for each $i < d$, $\theta^{(i')} = (\theta_1, \ldots, \theta_i, \theta'_{i+1}, \ldots, \theta'_d)$ and $\theta^{(d')} = \theta$. As noted in Schervish and Carlin (1992), if g_i is the joint density of the observations sampled at iteration i, then the joint density of the observations sampled at the next iteration (g_{i+1}) is given by

$$\int K(\theta', \theta)g_i(\theta')d\lambda(\theta'), \quad K(\theta', \theta) = \prod_{i=1}^{d} p^{(i)}(\theta^{(i')}) \tag{6.3.1}$$

[see also Tanner and Wong (1987) and Liu, Wong and Kong (1994, 1995)]. One may approximate the integral in (6.3.1) via the method of Monte Carlo. In particular, given the observations $\theta^1, \theta^2, \ldots, \theta^m$, use the Monte Carlo sum

$$\frac{1}{m} \sum_{j=1}^{m} K(\theta^j, \theta) \tag{6.3.2}$$

to approximate $g_{i+1}(\theta)$. Ritter and Tanner (1992) suggest using θ values from independent chains. One may also use *successive* θ values from one chain to construct the Monte Carlo sum (6.3.2). Note that construction of (6.3.2) requires the normalizing constants (or good approximations to the normalizing constants) for the conditional distributions. Also note that we are examining, through $p(\theta_k \mid \theta_1, \ldots, \theta_{k-1}, \theta_{k+1}, \ldots, \theta_d, Y)$, the first component of each θ vector along with components $2 - d$ of the other $m - 1$ θ vectors, the first and second components of each θ vector along with components $3 - d$ of the other $m - 1$ θ vectors, etc., thereby expanding the coverage of the parameter space. The effort in constructing (6.3.2) will yield useful information regarding the state of the Markov chain vis-à-vis the equilibrium distribution.

To illustrate this convergence diagnostic, consider the "Witch's Hat" distribution presented in Matthews (1991). The posterior under consideration is proportional to a mixture of a multivariate normal distribution and uniform distribution on the open d-dimensional hypercube $(0, 1)^d$ C:

$$(1 - \delta) \left(\frac{1}{\sqrt{2\pi}\sigma} \right)^d e^{-\frac{1}{2} \sum_{i=1}^{d} \left(\frac{y_i - \theta_i}{\sigma} \right)^2} + \delta I_{(Y \varepsilon C)}. \qquad (6.3.3)$$

We have chosen $\delta = 10^{-11}, \sigma = 0.03, d = 9$ and $Y = (0.9, 0.9, 0.9, 0.9, 0.9, 0.9, 0.9, 0.9, 0.9, 0.9)$. A cursory examination of the posterior reveals a spike centered at Y, with a flat "brim" extending out to the boundary of the unit hypercube. We proceed under the assumption that the posterior at hand is analytically complicated, thereby not allowing for an easy recognition of the location of or number of spikes. Clearly, in a situation where determining the number and location of the modes is straightforward, one would focus attention around these points, possibly along the lines suggested by Gelman and Rubin (1992).

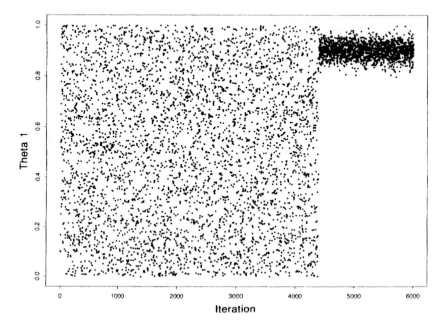

FIGURE 6.3. θ_1 vs. iteration.

FIGURE 6.4. Weights vs. iteration.

Figure 6.3 (see Cui et al., 1992) presents a history of the θ_1 marginal, starting from the point (0.1, 0.1, 0.1, 0.1, 0.1, 0.1, 0.1, 0.1, 0.1). The plots for the other marginals are quite similar. As can be seen from the plot, the Markov chain wanders about the hypercube until iteration 4400 or so, at which point it locates the mode. A run shorter than 4400 iterations would not have detected the spike.

Figure 6.4 presents a plot of $w(\theta^{(i)})$ vs. iteration i, $i = 1, 270$. The weights in this plot are based on the first 270 successive (9-dimensional) points in the chain. The 270 weights in the plot were standardized to have mean zero and unit standard deviation. As can be seen in the plot, all 270 weights are equal, with the exception of two weights.

The reason for the outlying weights is easily explained. The conditional densities $p(\theta_k \mid \theta_1, \ldots, \theta_{k-1}, \theta_{k+1}, \ldots, \theta_d, Y)$ used in computing K in (6.3.1) are proportional to (6.3.3) for specified θ_k and $0 < \theta_i < 1$. When θ_i is "far" from y_i, $i \neq k$, $p(\theta_k \mid \theta_1, \ldots, \theta_{k-1}, \theta_{k+1}, \ldots, \theta_d, Y)$ is virtually equal to unity, independent of the value of θ_k. When θ_i is "near" y_i, $i \neq k$, $p(\theta_k \mid \theta_1, \ldots, \theta_{k-1}, \theta_{k+1}, \ldots, \theta_d, Y)$ (as a function of θ_k) follows the normal curve. The outlying weight points noted in

Figure 6.2 stem from the fact that in three of the 270 products in (6.3.2), which are averaged to compute g_{i+1}, one of the terms $p(\theta_k \mid \theta_1, \ldots, \theta_{k-1}, \theta_{k+1}, \ldots, \theta_d, Y)$ has seven of the eight θ_i's within three σ's of 0.9, $i \neq k$, with θ_k "far" from 0.9. (For future reference, call these components θ_i^*, $i \neq k$.) This term is nearly zero, thereby causing the entire product of terms in K to be small. Thus, rather than averaging 270 products equal to unity to compute g_{i+1}, the average was diminished by the three near-zero products, leading to the outlying weight.

As a followup to the investigation of the cause of the two outliers, the maximizer (θ_k^*) of $p(\theta_k \mid \theta_1, \ldots, \theta_{k-1}, \theta_{k+1}, \ldots, \theta_d, Y)$ was located, where the θ_i's were set equal to the $\theta_i^* (i \neq k)$ values identified in the previous paragraph. A chain was started from this point. Figure 6.5 presents a plot of the history of the θ_1 marginal for this path; the other marginals are similar. As indicated in Figure 6.5, the chain moved immediately into the neighborhood of the spike.

In this example, we see how a careful examination of the outlying weights helps to locate the spike much earlier than the 4400 iterations required by the original chain. Of course, this Gibbs stopper technique is not infallible. Had we considered

FIGURE 6.5. θ_1 vs. iteration.

only the first 100 points in the Markov chain, we would have missed the outliers. Similarly, one would expect that smaller values of σ and higher values of d would require more terms in (6.3.2)—though both of these modifications would probably increase the run time of the chain as well.

Roberts (1992) essentially suggests calculating the average of the Gibbs stopper weights (his Λ_n). He formally shows that the expectation of Λ_n goes to unity as the chain moves to the equilibrium distribution. Roberts (1992) also calculates the asymptotic variance of Λ_n (as the number of iterations goes to infinity).

Additional weight plots based on nonoverlapping segments (iterations 271–810, 811–1850, 1851–4010, etc.) of the original Markov chain are presented in Figures 6.6–6.9. One does not see in the first three plots of this series a degeneration of the distribution of the weights about a spike, as would be expected if the chain was in equilibrium. This provides further indication to the data analyst of an anomaly regarding convergence of the chain. These plots clearly highlight the slowly mixing nature of the chain—a valuable diagnostic for the data analyst. The final plot in the series highlights the fact that the equilibrium distribution has two

FIGURE 6.6. Weights vs. iteration.

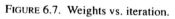

FIGURE 6.7. Weights vs. iteration.

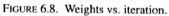

FIGURE 6.8. Weights vs. iteration.

FIGURE 6.9. Weights vs. iteration.

distinct domains of attraction—the "brim" and the "spike". See Section 6.4.3 for further illustration of Gibbs stopper methods.

It is also useful in assessing convergence via the Gibbs stopper to plot the average and standard deviation of the weights in successive batches of (possibly) increasing size. To illustrate this diagnostic, consider using the Gibbs sampler to draw from the trivariate normal with mean $(0, 0, 0)$ and variance–covariance matrix:

$$\begin{bmatrix} 1 & 4.5 & 9 \\ 4.5 & 25 & 49 \\ 9 & 49 & 100 \end{bmatrix}$$

(see Cowles and Carlin, 1996). In other words, rather than draw directly from this distribution (see Exercise 4 of Chapter 1), the Gibbs sampler was run where the equilibrium distribution in this trivariate normal with high correlations of 0.9, 0.9 and 0.98. The full conditionals of the Gibbs sampler were the conditional univariate normals.

Figures 6.10 and 6.11 are based on one long chain of Gibbs sampler iterates. Figure 6.10 represents the batch mean of the weights for 10 successive batches

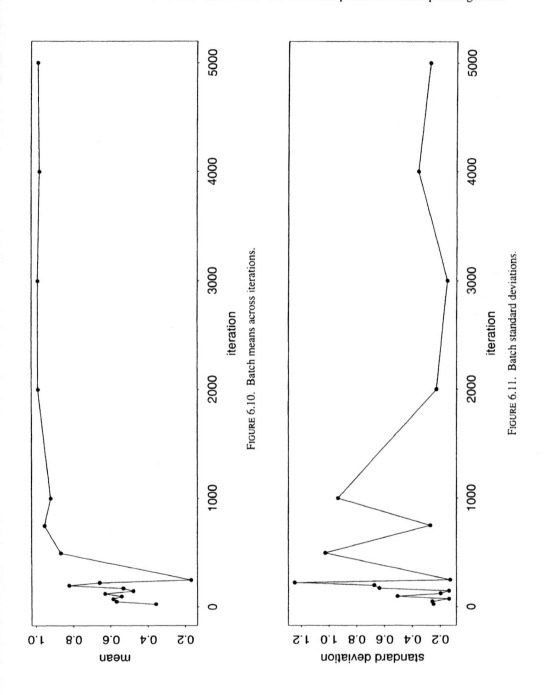

FIGURE 6.10. Batch means across iterations.

FIGURE 6.11. Batch standard deviations.

of size 25, followed by three batches of size 250 and then four batches of size 1000. Figure 6.11 presents the corresponding information for the batch standard deviation. Within each batch, the iterates were used to compute g_i and then we computed the weights over the points in the batch.

Looking at the results for the first 10 batches of size 25, it is not clear what is happening. Is the fluctuation in the batch means due to the possibility that the algorithm has "converged" but m is too small, is it due to the algorithm not yet having "converged" (but m is large enough), or is it due to the possibility that the algorithm has not converged *and* m is too small? This ambiguity suggests that further iteration with a larger batch size is needed. Increasing the batch size to 250 provides a more consistent picture in terms of the batch means, though the within-batch standard deviations are still quite high—about as high as the $m = 25$ case. Again, the results suggest that further iteration with a larger batch size is in order. The batch size was increased to 1000, though a batch size of 500 may have been sufficient. Examining the plot, we see that the batch means for $m = 1000$ are slightly higher than for the 250 case—though there is consistency across the $m = 1000$ batches for both means and standard deviations. The standard deviation of the weights dropped in magnitude (compared to the $m = 250$ case) and remained low.

The batch mean and batch standard deviation plots suggest that in this case a tour of length in excess of 250 iterations but less than 1000 seems to be needed to capture adequately the nature of the posterior distribution. The plots also suggest that the point at iteration 1000 may be a good place to start the tour, but earlier points may have worked as well. In addition, we learn from the batch mean plot that the normalizing constant on the objective function is estimated to be close to the true value of 1.0.

6.3.2. Control Variates

Liu and Liu (1993) have suggested using the following control variates when m chains have been generated:

$$U^{(i,j,t)} = \frac{p(\theta^{(j,t)} \mid Y)}{p(\theta^{(i,t)} \mid Y)} \frac{K(\theta^{(j,t-1)}, \theta^{(i,t)})}{K(\theta^{(j,t-1)}, \theta^{(j,t)})},$$

where $\theta^{(j,t)}$ is the tth value in chain j, $\theta^{(i,t)}$ is the corresponding value in chain i, for $i \neq j$, $i, j = 1, \ldots, m$ and $K(\cdot, \cdot)$ is defined in 6.3.1. They prove that the expected value of $U^{(i,j,t)}$ is equal to

$$\text{var}_p \left(\frac{g_t(\theta)}{p(\theta \mid Y)} \right) + 1,$$

where g_t is the current approximation to the posterior. Hence, each $U^{(i,j,t)}$ is an unbiased estimate of the distance between $g_t(\theta)$ and $p(\theta \mid Y)$. In their paper, Liu and Liu (1993) plot the average of the logarithm of $U^{(i,j,t)}$ (over the $m(m-1)$ pairs of chains) vs. t. As $g_t(\theta)$ converges to $p(\theta \mid Y)$, the resulting curve should fluctuate about a line parallel to the x-axis.

6.3.3. Alternative Methods

Other approaches for assessing convergence have been proposed. Gelman and Rubin (1992) consider the ratio of between-chain (across m chains) variation to within-chain (up to iteration n) variation. If this ratio is high, then there may be a reason to believe that further iteration will be of value. Formally, they compute:

$$\sqrt{\hat{R}} = \sqrt{\left(\frac{n-1}{n} + \frac{(m+1)}{mn}\frac{B}{W}\right)\frac{df}{df-2}} = \sqrt{\frac{\hat{V}}{W}\frac{df}{df-2}},$$

where $B/n =$ the variance between the m chain means, $\bar{x}_{i.}$, each based on n values of the scalar x; $B/n = \Sigma_{i=1}^{m}(\bar{x}_{i.} - \bar{x}_{..})^2/(m-1)$; $W =$ the average of the m within-chain variances, s_i^2, each based on $n-1$ degrees of freedom, i.e. $W = \Sigma_{i=1}^{m} s_i^2/m$; and

$$df = 2\hat{V}^2/\widehat{\text{var}}(\hat{V}),$$

where

$$\hat{V} = \frac{n-1}{n}W + \left(\frac{m+1}{mn}\right)B$$

and

$$\widehat{\text{var}}(\hat{V}) = \left(\frac{n-1}{n}\right)^2\frac{1}{m}\widehat{\text{var}}(s_i^2) + \left(\frac{m+1}{mn}\right)^2\frac{2}{m-1}B^2$$
$$+ 2\frac{(m-1)(n-1)}{mn^2}\cdot\frac{n}{m}\cdot\left[\widehat{\text{cov}}(s_i^2, \bar{x}_{i.}^2) - 2\bar{x}_{..}\,\widehat{\text{cov}}(s_i^2, \bar{x}_{i.})\right],$$

and the estimated variances and covariances are obtained from the m sample values of $\bar{x}_{i.}$ and s_i^2. When $\hat{R} \approx 1$ for all scalar estimands of interest, Gelman and Rubin (1992) conclude that one can terminate iteration. Code for this procedure can be found at http://lib.stat.cmu.edu/S/itsim

The idea of comparing between-chain variation to within-chain variation was presented in Tanner and Wong (1987) in the context of data augmentation. This approach may not work well if the chains are trapped within the same subregion. Applications of this methodology are given in Gelman and Rubin (1992).

Raftery and Lewis (1992) discuss the problem of how many times to iterate the Gibbs sampler. In particular, suppose that one wishes to estimate a particular quartile of the posterior distribution $p(U \le \mu \mid Y)$ to within $\pm r$ with probability s, where U is a function of the parameter vector θ. Raftery and Lewis (1992) present formulas (for the single chain case) for the number of initial M iterations that are to be discarded, and for a further N iterations of which every kth is stored.

Suppose that one calculates U_t for each iteration t and then forms $Z_t = \delta(U_t \leq \mu)$, where $\delta(\cdot)$ is the indicator function. $\{Z_t\}$ is a binary 0–1 process. Consider the new process $\{Z_t^{(k)}\}$, where $Z_t^{(k)} = Z_{1+(t-1)k}$, and let

$$\begin{bmatrix} 1 - \alpha, & \alpha \\ \beta, & 1 - \beta \end{bmatrix}$$

be the transition matrix for $(Z_t^{(k)})$. The number of "burn-in" iterations to be discarded is equal to mk, where

$$m = \frac{\log\left(\dfrac{\varepsilon(\alpha + \beta)}{\max(\alpha, \beta)}\right)}{\log(1 - \alpha - \beta)}$$

and ε is the required distance between $p(z_m^{(k)} = i \mid z_0^{(k)} = j)$ and the equilibrium probability. The formula for N is kn, where

$$n = \frac{\alpha\beta(2 - \alpha - \beta)}{(\alpha + \beta)^3} \left[\frac{\Phi\left(\frac{1}{2}(1 + s)\right)}{r}\right]^2$$

and $\Phi(\cdot)$ is the standard normal cdf. To implement this idea, one needs an initial run to estimate α and β. A poor estimate of α and β may result if the chain is trapped in a subregion of the parameter space.

Geweke (1992) uses methods from spectral analysis to assess convergence of the Gibbs sampler. Suppose that the goal is to estimate the posterior mean of the function $g(\theta)$, i.e. $\int g(\theta)p(\theta \mid Y)d\theta$. We can estimate this posterior mean based on n successive iterations of the Gibbs sampler via

$$\bar{g}_n = \frac{\sum_{i=1}^n g(\theta^i)}{n},$$

with asymptotic variance $S_g(0)/n$. This assumes the existence of a spectral density $S_g(\omega)$ for the time series $g(\theta_1), \ldots, g(\theta_n)$ that has no discontinuities at the frequency 0. Geweke (1992) suggests that one take the difference between $\bar{g}_{n_a}$ (based on the first n_a iterations) and $\bar{g}_{n_b}$ (based on the last n_b iterations) and divide by the asymptotic standard error of the difference (using spectral density estimates for the two pieces of the time series), i.e.

$$\frac{\bar{g}_{n_a} - \bar{g}_{n_b}}{\sqrt{\dfrac{\hat{S}_g^a(0)}{n_a} + \dfrac{\hat{S}_g^b(0)}{n_b}}},$$

where $\hat{S}_g^b(0)$, for example, is an estimate of the spectral density for the last part of the time series. If the chain is stationary, then the mean of the first part of the chain should be similar to the mean of the last part of the chain. Assuming that the ratios n_a/n and n_b/n are held fixed and $n_a + n_b < n$, then asymptotically (with n) the statistic has a standard normal distribution if the time series is stationary. Geweke (1992) estimates $S_g(0)$ based on the periodogram of $g(\theta_1), \ldots, g(\theta_n)$ using a Daniell window of width $2\pi/.3\sqrt{n}$ (Hannan, 1977). Geweke (1992) suggests that $n_a = .1n$ and $n_b = .5n$. Note that if the chain is trapped in a local mode, this technique may still indicate convergence.

Cowles and Carlin (1996) present a comparative review of convergence diagnostics for MCMC algorithms. Code to implement some of these diagnostics is available via anonymous ftp along with the BUGS program referenced at the end of Section 6.1.

6.4 The Griddy Gibbs Sampler

In many practical situations, the distribution $p(\theta_i \mid \theta_1, \ldots \theta_{i-1}, \theta_{i+1}, \ldots \theta_d, Y)$ is univariate. When it is difficult to sample directly from $p(\theta_i \mid \theta_1, \ldots \theta_{i-1}, \theta_{i+1}, \ldots \theta_d, Y)$, one may form a simple approximation to the inverse cdf based on the evaluation of $p(\theta_i \mid \theta_1, \ldots \theta_{i-1}, \theta_{i+1}, \ldots \theta_d, Y)$ on a grid of points. Ritter and Tanner (1992) consider the following algorithm:

Step 1. Evaluate $p(\theta_i \mid \theta_1, \ldots \theta_{i-1}, \theta_{i+1}, \ldots \theta_d, Y)$ at $\theta_i = \phi_1, \phi_2, \ldots, \phi_n$, to obtain $w_1, w_2, \ldots, w_n$.

Step 2. Use $w_1, w_2, \ldots, w_n$ to obtain an approximation to the inverse cdf of $p(\theta_i \mid \theta_1, \ldots \theta_{i-1}, \theta_{i+1}, \ldots \theta_d, Y)$.

Step 3. Sample a Uniform [0,1] deviate and transform the observation via the approximate inverse cdf.

Remark 6.4.1. The function $p(\theta_i \mid \theta_1, \ldots \theta_{i-1}, \theta_{i+1}, \ldots \theta_d, Y)$ need only be known up to a proportionality constant, since the normalization can be obtained from the $w_1, w_2, \ldots, w_n$ directly.

Remark 6.4.2. The grid points need not be uniformly spaced. In fact, good grids put more points in neighborhoods of high mass and fewer points in neighborhoods of low mass. One approach to address this goal is to construct the grid so that the mass under the current approximation to the marginal distribution between successive grid points is approximately constant. This point is discussed in more detail below.

Remark 6.4.3. The number of points in the grid need not be constant over the iterations of the Gibbs sampler. At early iterations, n may be small. As the algorithm iterates toward the joint distribution, n can be increased. In practice, when the algorithm appears to have stabilized with $n = n^*$ grid points, at the next iteration one may use a finer grid e.g. $n = 2n^*$ or $4n^*$. The Gibbs stopper may also be used to regulate n.

Remark 6.4.4. Simple approximations to the inverse cdf are:

a. Piecewise constant corresponding to a discrete distribution for $\phi_1, \ldots, \phi_n$ with probabilities $p(\phi_i) = w_i / \sum_{j=1}^{n} w_j$.
b. Piecewise linear corresponding to a piecewise uniform distribution on the interval $[a_i, a_{i+1}]$, $i = 1, \ldots, n$, with $\phi_i \in [a_i, a_{i+1}]$ and density $f_i = w_i / \sum_{j=1}^{n} \omega_j$, where $\omega_i = w_i(a_{i+1} - a_i)$. Typically, ϕ_i is centered in the interval $[a_i, a_{i+1}]$.

More sophisticated approximations may be based on piecewise quadratic interpolation or higher-order splines. In general, when the conditional marginal is easy to evaluate, one may wish to use a simpler approximation to the inverse cdf and use a finer grid. When the conditional marginal is more difficult to evaluate, one may wish to use a coarser grid but a more clever approximation. Tierney (1994) discusses how this algorithm may be made "exact" using a Metropolis step. See also Dellaportas (1995).

Remark 6.4.5. Over an unbounded interval, it is important to compare w_1 and w_n to $M = \max(w_1, \ldots, w_n)$. If w_1 is greater than $f \cdot M$ $(0 < f < 1)$, then the grid must be augmented with points to the left of ϕ_1. If w_n is greater than $f \cdot M$, then the grid must be augmented with points to the right of ϕ_n. In practice, it is sufficient to force the grid to grow if at either end of the grid the density is more than 10% of the maximum value. If, however, interest centers on the tails of the distribution, then f should be smaller than 0.1. We will refer to this device as the *grid-grower*.

Several alternative methods are available for handling the situation where it is difficult to sample directly from the conditionals. These methods tend to rely on importance sampling or acceptance/rejection approaches. Wei and Tanner (1990c) presented an importance sampling approach; see Section 5.9.1. Carlin and Gelfand (1991) and Zeger and Karim (1991) presented rejection/acceptance modifications to the Gibbs sampler. (See Section 6.2.3 for the approach of Zeger and Karim). Also of note is the work by Gilks and Wild (1992), who used tangent and secant approximations above and below the log-posterior to develop an acceptance/rejection scheme. Acceptance/rejection methods are exact in the sense that they produce samples from the required distribution. Possible drawbacks of these methods include a low acceptance rate, the need to know the normalizing constant of the conditional distribution, and possible restrictions on the distribution (e.g. log-concavity). As illustrated in Section 6.4.3, nonlinear regression problems may lead to conditionals that are not log-concave and may in fact be multimodal. Moreover, in nonlinear regression the conditionals, typically, are known only up to a multiplicative constant. Both importance sampling and rejection/acceptance algorithms require a higher degree of programming sophistication on the part of the data analyst and thereby detract from the conceptual simplicity and appeal of the Gibbs sampler. Specification of an importance sampling function that is easy to sample from, yet provides a "good match" to the density of interest, may require the specification of several tuning constants. In the context of rejection/acceptance, specification of tuning constants is required to ensure that the importance function, modulo a multiplicative constant, dominates the density of interest *everywhere* (or at least over a region of high content; e.g. 95%. An important alternative approach based on the Metropolis algorithm is discussed in Section 6.5.3.

6.4.1. Example

For the genetic linkage model with data (14, 0, 1, 5), the distribution of the latent data (x_2) given θ and Y, $p(x_2 \mid \theta, Y)$, is the binomial distribution with parameters 14 and $\theta/(\theta + 2)$. The augmented posterior, $p(\theta \mid x_2, Y)$, is (under a flat prior) proportional to

$$\theta^{x_2+x_5}(1 - \theta)^{x_3+x_4}. \tag{6.4.1}$$

To implement the Gibbs sampler, one would draw x_2^* from $p(x_2 \mid \theta^*, Y)$, where θ^* is the current value for θ. The next step would be to sample a new θ^* from $p(\theta \mid x_2^*, Y)$. These two steps would then be iterated. Independent replicates of this path could be constructed to realize an independent sample of θ values from $p(\theta \mid Y)$. For the purpose of this example, the fact that the augmented posterior is a beta distribution is ignored. We only assume that one can calculate (6.4.1) at any value of θ. Let m denote the number of such paths.

The griddy Gibbs sampler based on a linear approximation of the inverse cdf was initially implemented with a five-point grid on [0,1]. The algorithm was allowed to run for five iterations, at which point the number of grid points was increased to ten. The algorithm continued for five more iterations, at which point the number of grid points was increased to 20. The final five iterations used 40 grid points.

Figure 6.12 summarizes the history of this process by plotting the upper, middle and lower quartile for θ vs. iteration number. (Three hundred independent griddy Gibbs sampler paths were created to obtain the independent θ values.) As can be seen from Figure 6.12, the median appears to have stabilized quite quickly with a small number of grid points. However, the lower tail does require further iteration and a finer mesh to stabilize.

As can be seen in Figure 6.13, the histograms of the θ values after 15 and 20 iterations (with $n = 20$ and 40, respectively) are quite similar, and both are congruent with the true posterior. Thus, a grid with 20 equally spaced points appears to capture the shape of $p(\theta \mid x_2, Y)$.

6.4.2. Adaptive Grid/Grid Grower

In the previous section, a grid with uniform gap size was used at each iteration. Clearly, a grid that has more points in a region of high mass and fewer points in a region of less mass is preferable.

One approach to address this goal is to make use of the empirical cdf($\hat{F}$) of the X_i values (obtained from the m replicates of the griddy Gibbs path) from the

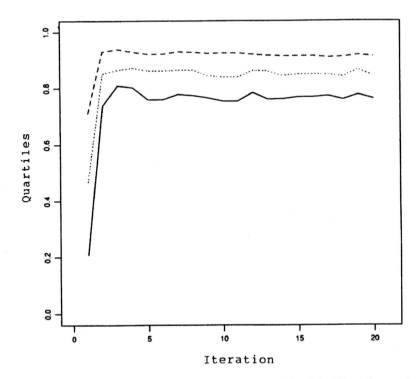

FIGURE 6.12. Sample quartiles vs. iteration. $n = 5, 10, 20, 40$ for 5 iterations each.

current iteration. The new n grid points $\phi_1, \ldots, \phi_n$ can be obtained by computing the empirical quantiles $\hat{F}^{-1}(\zeta_i)$ for preselected values $0 < \zeta_1 < \cdots < \zeta_n < 1$. This grid is referred to as the *adaptive grid*.

Remark 6.4.6. Usually, the ζ_i are chosen to be equidistant. This would form a grid that is dense in regions of high mass and sparse in regions of low mass.

Remark 6.4.7. It is not efficient to modify the grid at each iteration. Rather, one can branch to this adaptive-grid method every j iterations. In the example below, $j = 5$.

Remark 6.4.8. One may use the adaptive-grid idea to increase the number of points as well as to specify where to place the points, i.e. n may be some multiple (other than one) of the number of grid points in the previous iteration.

Remark 6.4.9. The adaptive grid may be used in conjunction with the grid grower introduced above.

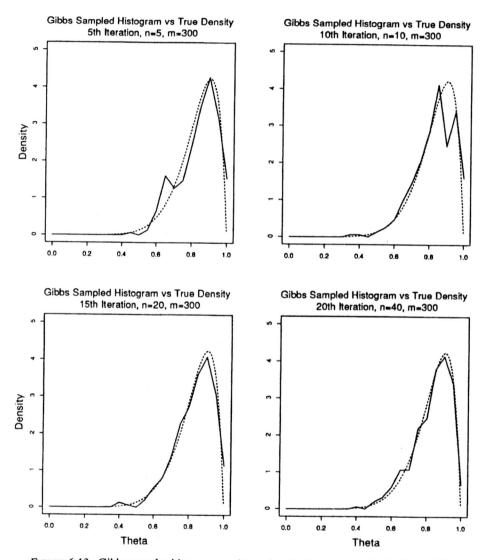

FIGURE 6.13. Gibbs sampler histogram and true density for iterations 5, 10, 15 and 20.

1. Every i_0 iterations compute $\hat{F}$ and derive a grid size n from it, as described above. This grid is called the *global grid*.
2. For the first of the d conditional distributions, augment the global grid, if necessary, using the grid grower. The corresponding grid will be called the *local grid*. Use this local grid for the next d conditional distribution and augment, if necessary, using the grid grower.

If the global grid is far from the hump of the specific marginal conditional distribution, then the grid grower will correct the problem at the expense of function evaluations. Having captured the important part of the density, the grid grower would tend not to be used.

6.4.3. Nonlinear Regression

As a second example, we consider the biochemical oxygen demand (BOD) data from Marske (1967), which was analyzed by Bates and Watts (1988). To determine the biochemical oxygen demand, a sample of stream water was taken, injected with soluble organic matter, inorganic nutrients and dissolved oxygen, and subdivided into BOD bottles. Each bottle was inoculated with a mixed culture of microorganisms, sealed and incubated at constant temperature. The bottles were opened periodically and analyzed for dissolved oxygen concentration, from which the BOD was calculated in milligrams per liter. The values shown in Table 6.3 are the averages of two analyses on each bottle.

A nonlinear model was derived based on exponential decay with a fixed rate constant (Bates and Watts, 1988) as

$$y_i = \theta_1[1 - \exp(-\theta_2 x_i)] + \varepsilon_i,$$

where y is the biochemical oxygen demand at time x. For independent normal errors with constant variance σ^2, the likelihood for the BOD problem is

TABLE 6.3. Biochemical
Oxygen Demand vs. Time.

Time (days)	BOD (mg/I)
1	8.3
2	10.3
3	19.0
4	16.0
5	15.6
7	19.8

Source: Marske, 1967.

$$L(\theta_1, \theta_2, \sigma^2 \mid Y) = \exp\left[-n \log \sigma - \frac{1}{2} \frac{S(\theta_1, \theta_2)}{\sigma^2}\right],$$

where $S(\theta_1, \theta_2)$ is the sum of squared residuals.

Assuming a flat prior and integrating out σ^2 results in the following (improper) posterior:

$$p(\theta_1, \theta_2 \mid Y) \propto [S(\theta_1, \theta_2)]^{-[n/2-1]}.$$

Figure 6.14 presents the likelihood contours. Contours above the approximate 93% F level are open with respect to θ_2. Note that in this region the model is insensitive to changes in θ_2. Moreover, for values of θ_2 less than 0, θ_1 is negative, thus realizing a fold in the 99.9% contour.

For the purpose of the present analysis, a flat (strictly positive) prior for (θ_1, θ_2) is adopted over the region $[-20,50] \times [-2,6]$. Outside this region, the prior for (θ_1, θ_2) is zero. This amounts to using the restriction that $p(\theta_1, \theta_2 \mid Y)$ be normalized on the domain $[-20,50] \times [-2,6]$ to integrate to unity. Under this restriction, the posterior is bimodal, with the main mode at (19,0.5) and a minor mode at the lower-left corner.

We illustrate in Figure 6.15 the griddy Gibbs sampler with adaptive grids. Grids of 20 points were used spanning the ranges $[-20,50]$ and $[-2,6]$. The algorithm was initialized with a uniform sample on the domain and 500 paths of griddy Gibbs sampler were run. Figure 6.15 presents the scatter plot of (θ_1, θ_2) for selected iter-

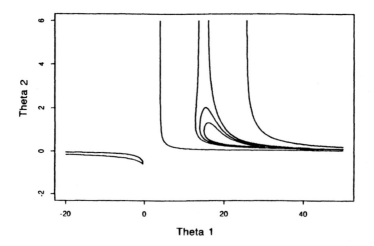

FIGURE 6.14. 80%, 90%, 95% and 99.9%. Normal likelihood contours for θ_1, and θ_2 labelled by approximate frequency coverage using the F statistic.

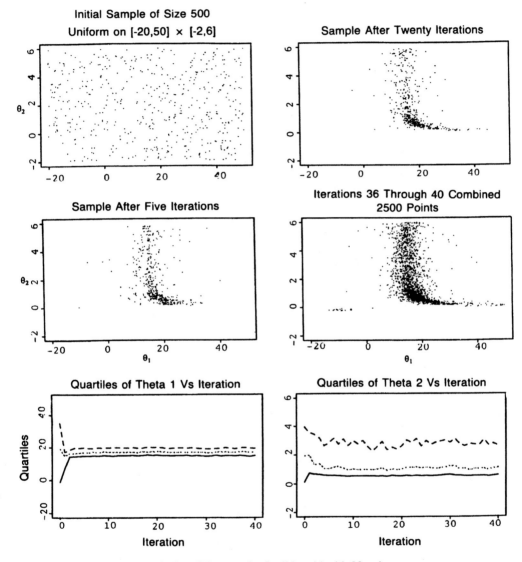

FIGURE 6.15. BOD example, flexible grid with 20 points.

ations as well as the quartiles of θ_1 and θ_2 as functions of the number of iterations. Note that the quartiles appear to stabilize quite early.

Figure 6.16 compares the θ_2 marginals obtained (1) by numerical integration and (2) by forming the histogram of the θ_2 values of iterations 36–40 combined. From this plot, it can be seen that the griddy Gibbs sampler has recovered this marginal.

Finally, Figure 6.17 presents the Gibbs stopper weights for iterations 1, 5, 35 and 40. As can be seen from these plots, a major component of the decrease in variation occurred by the fifth iteration. Beyond the fifth iteration, one may wish to increase the number of points in the grid or the number of chains to improve the precision of the estimate. Had all four distributions been similar in shape to the top plot, one would tune the number of points in the grid or the number of chains to push the approximation toward the true joint distribution.

6.4.4. Cox Model

As a third example, we consider data taken from Glasser (1965). The response is the survival time in days for subjects with primary lung tumors. Two covariates are considered: the age of the patient and the performance status. Of the 16 cases, six are right-censored. We model these data using the Cox proportional hazards model

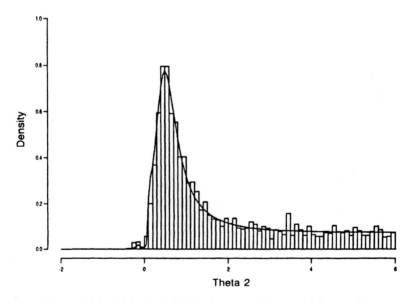

FIGURE 6.16. θ_2 Marginal. Solid—integrated; Bars—Gibbs sample.

FIGURE 6.17. Stopper weights for flexible grid with $n = 20$.

(Lawless, 1982, Chapter 7). The partial likelihood is used as the basis of inference for this model. For the Glasser data, the mode of the partial likelihood is located at -0.217 (performance) and at -0.005 (age), with corresponding estimated standard errors of 0.162 and 0.042, respectively. The results after 20 iterations of the griddy Gibbs sampler (applied to the partial likelihood with a flat prior) are quite similar. The average parameter value based on 300 independent chains is -0.208 for performance and -0.005 for age, and the associated standard deviations are

0.162 and 0.043, respectively. The estimated survivor function corresponding to a performance status of 7 and an age of 51, at 141 days, is 0.625 and 0.641, respectively, for the modal partial likelihood and the griddy Gibbs approaches. The estimated standard error based on the griddy Gibbs approach is 0.085. The griddy Gibbs point estimate for the survivor function (at $t = 141$) was obtained by averaging the survivor function $S(t \mid \mathbf{x}) = S_0(t)^{\exp(\mathbf{x}\beta)}$ over the 300 independent values of the parameter vector (β), where $\mathbf{x}$ is the known covariate vector. (The dependency of the baseline survivor function ($S_0(t)$) on the parameter vector is given by Lawless (1982, page 361).) The associated estimated standard error was obtained by computing the standard deviation of these 300 survivor function values. Clayton (1991) and Sinha (1993) apply data augmentation and Gibbs sampler algorithms to the related frailty models.

6.5 The Metropolis Algorithm

The Metropolis algorithm (Metropolis et al. 1953) was developed to investigate the equilibrium properties of large systems of particles such as electrons in an atom. This algorithm has been extensively used in the statistical physics literature (Hammersley and Handscomb, 1964) to simulate complex systems. Hastings (1970) suggests a generalization of the Metropolis algorithm. He illustrates how to use this algorithm to simulate Poisson and normal deviates, as well as random orthogonal matrices. In this section, we show how one can use the Metropolis algorithm to construct a Markov chain with equilibrium distribution $\pi(x)$. However, before presenting the algorithm, we briefly review some elementary ideas related to discrete-space Markov chains.

6.5.1. Elements of Discrete-space Markov Chains

Assume that we have a set of states, $S = \{s_1, s_2, \ldots, s_d\}$. At time 0, the process begins in one of these states, and the process moves from sate s_i to state s_j at unit time intervals. For a Markov chain, the status of the chain at time $n + 1$ depends only on its status at time n. Hence, the transition probability p_{ij} that the state moves from s_i to s_j may be organized in a $d \times d$ matrix, where $p_{ij} \geq 0$ and $\sum_j p_{ij} = 1$ for all rows i.

As an illustration, we consider an example from Snell (1988). In the Land of Oz there are three kinds of weather: rain, nice and snow. The transition probabilities are

$$P = \begin{array}{c} R \\ N \\ S \end{array} \begin{pmatrix} 0.5 & 0.25 & 0.25 \\ 0.5 & 0 & 0.5 \\ 0.25 & 0.25 & 0.5 \end{pmatrix}.$$

Suppose that on day 0 the weather is nice (N) and let, for example, $p(X_i = S)$ be the probability of snowy weather on day i. The probabilities of the three states

(R, N, S) on day 1 would then follow from the transition matrix to be

$$(p(X_1 = R), p(X_1 = N), p(X_1 = S))$$
$$= (w_1^{(1)}, w_2^{(1)}, w_3^{(1)}) = w^{(1)} = (0.5, 0, 0.5).$$

What are the corresponding probabilities for day 2? In general, we compute the probabilities for day $n + 1$ by noting that

$$w_j^{(n+1)} = p(X_{n+1} = s_j)$$
$$= \sum_k p(X_n = s_k, X_{n+1} = s_j)$$
$$= \sum_k p(X_n = s_k)p(X_{n+1} = s_j \mid X_n = s_k)$$
$$= \sum_k w_k^{(n)} p_{kj}.$$

In matrix notation, we have $w^{(n+1)} = w^{(n)} P = w^{(0)} P^n$. In this way, starting from a nice day, i.e. $p(X_0 = N) = 1$, we have

$$w^{(1)} = (0.5, 0, 0.5)$$
$$w^{(2)} = (0.375, 0.25, 0.375)$$
$$w^{(3)} = (0.406, 0.188, 0.406)$$
$$w^{(4)} = (0.398, 0.203, 0.398)$$
$$w^{(5)} = (0.4, 0.2, 0.4)$$
$$w^{(6)} = (0.4, 0.2, 0.4).$$

Similarly, if $p(X_0 = R) = 1$ (i.e. we started on a rainy day) or $p(X_0 = S) = 1$ (i.e. day 0 was snowy), we would have $w^{(6)} = (0.4, 0.2, 0.4)$. Thus, if we were to simulate the behavior of this chain, we would find in the long run that 40% of the days would be in state R, 20% in state N and 40% in state S, irrespective of the starting point of the chain. The vector $(0.4, 0.2, 0.4)$ is called the *equilibrium distribution*.

More generally, a Markov chain with transition matrix P will have an equilibrium distribution π iff $\pi = \pi P$. This chain is a *reversible chain* iff $\pi_i p_{ij} = \pi_j p_{ji}$ for all $i \neq j$. Note that $\pi_i p_{ij} = \pi_j p_{ji}$ implies $\pi = \pi P$, since

$$(\pi P)_j = \sum_i \pi_i p_{ij} = \sum_i \pi_j p_{ji} = \pi_j \sum_i p_{ji} = \pi_j.$$

In this way, to sample from the equilibrium distribution π, we run a Markov chain with transition matrix P satisfying $\pi_i p_{ij} = \pi_j p_{ji}$ until the chain appears to have settled down to equilibrium. One of the contributions of Metropolis et al. (1953) is in detailing a general way of constructing P to realize a chain with equilibrium distribution π.

6.5.2. Metropolis' Method

First, we present the idea in the discrete case. The continuous case is taken up later. Let $Q = \{q_{ij}\}$ be a specified symmetric transition matrix. At a given step, randomly draw state s_j from the ith row of Q. With known probability α_{ij} move from s_i to s_j; otherwise, remain at step s_i. This construction defines a Markov chain with transition matrix $p_{ij} = \alpha_{ij}q_{ij}$ ($i \neq j$) and $p_{ii} = 1 - \Sigma_{j \neq i}p_{ij}$. Following Metropolis et al. (1953), we let

$$\alpha_{ij} = \begin{cases} 1 & \text{if } \pi_j/\pi_i \geq 1 \\ \pi_j/\pi_i & \text{if } \pi_j/\pi_i \leq 1 \end{cases}.$$

This chain is reversible, since

$$\pi_i p_{ij} = \pi_i \min\left\{1, \frac{\pi_j}{\pi_i}\right\} q_{ij}$$

$$= \min\{\pi_i, \pi_j\}q_{ij}$$

$$= \min\{\pi_i, \pi_j\}q_{ji}$$

$$= \pi_j p_{ji}.$$

The equilibrium distribution will be unique if Q is irreducible (Ripley, 1987). A sufficient condition for convergence (if π is not constant) is being able to move from any state to any other under Q (Ripley, 1987).

Barker (1965) takes $\alpha_{ij} = \pi_j/(\pi_i + \pi_j)$. Note that the resulting chain is reversible, since $\pi_i p_{ij} = [\pi_i \pi_j/(\pi_i + \pi_j)]q_{ij} = \pi_j p_{ji}$.

In the continuous case, where π is a density with respect to a measure μ and $f(x, y)$ is a symmetric transition probability function, [i.e. $f(x, y) = f(y, x)$], then the Metropolis algorithm is given by:

a. If the chain is currently at $X_n = x$, then generate a candidate value y^* for next location X_{n+1} from $f(x, y)$.
b. With probability

$$\alpha(x, y^*) = \min\left\{\frac{\pi(y^*)}{\pi(x)}, 1\right\}$$

accept the candidate value and move the chain to $X_{n+1} = y^*$. Otherwise, reject and let $X_{n+1} = x$. Thus, the Metropolis algorithm yields a series of *dependent* realizations forming a Markov chain with π as its equilibrium distribution. A key observation is that the Metropolis algorithm only requires that π be defined up to the normalizing constant since the constant drops out in the ratio $\pi(y^*)/\pi(x)$.

Tierney (1994) presents a number of suggestions for $f(x, y)$. If $f(x, y) = f(y - x) = f(x - y)$, then the chain is driven by a random-walk process. Possible candidates for f are the multivariate normal, multivariate t or split t (with a small degrees of freedom). In situations where the multivariate normal or multivariate t is used to generate candidate values, one would center the normal or the t at the current state of the chain x, with the variance–covariance matrix possibly equal to some multiple of the inverse information at the posterior mode. Müller (1993) discusses the choice of scale issue in detail. Besides presenting a range of hybrid strategies by cycling/mixing different chains, Tierney (1994) presents formal conditions for convergence, rates of convergence and limiting behavior of averages. Gelfand (1992) presents an analog to the Gibbs stopper for the Metropolis algorithm.

A generalization of the Metropolis algorithm due to Hastings (1970) takes

$$
\alpha(x, y) = \begin{cases} \min\left\{ \frac{\pi(y)q(y,x)}{\pi(x)q(x,y)}, 1 \right\} & \text{if } \pi(x)q(x, y) > 0 \\ 1 & \text{if } \pi(x)q(x, y) = 0 \end{cases},
$$

where $q(x, y)$ is an *arbitrary* transition probability function. Note that if q is symmetric, i.e. $q(x, y) = q(y, x)$, as would be the case in using a multivariate normal or multivariate t to drive the algorithm, then the Hastings algorithm reduces to the Metropolis algorithm. Hastings (1970) considers the case where $q(x, y) = q(y)$, which is closely related to importance sampling. Tierney (1994) calls these *independence chains*. See also Chib and Greenberg (1995).

EXAMPLE. *Genetic Linkage (Continued)*
We illustrate the Metropolis algorithm with the very simple genetic linkage model presented in Section 4.1. In the notation of this section, we have

$$
\pi(\theta) = (2 + \theta)^{125}(1 - \theta)^{38}\theta^{34} \propto p(\theta \mid Y), \qquad \theta \in [0, 1].
$$

Note that the normalizing constant for $p(\theta \mid Y)$ is not required for implementing the Metropolis algorithm. Also note that the Metropolis algorithm works with the observed posterior, rather than with $p(\theta \mid Z, Y)$ and with $p(Z \mid \theta, Y)$—as would be the case with data augmentation (Gibbs sampler).

In this example, candidate values were generated from three symmetric transition functions: the normal centered at the current value with standard deviation equal to 0.001, the normal centered at the current value with standard deviation equal to 0.05 and a corresponding normal with standard deviation equal to 0.12. Using the normal to drive the Metropolis algorithm leads to values outside the range [0,1]. In such instances, we recall that $\pi(\theta) = 0$ for values of θ outside the interval [0,1].

One long chain, starting from $\theta = 0.1$, was created for each driver, and the results are presented in Figures 6.18, 6.19 and 6.20. In each figure, the solid line represents the path of the chain, while the dotted/dashed line represents the proportion of accepted jumps for the Metropolis algorithm (computed over contiguous blocks of 50 iterations). The horizontal bar is located at $Y = 0.63$, the posterior mean. Note that the variance of the first driver is grossly underspecified, leading to a highly correlated sequence of values that require many iterations to "forget" the starting point and reach the equilibrium distribution. The acceptance rate for this chain tends to be quite high, suggesting that the chain is moving up the hill of

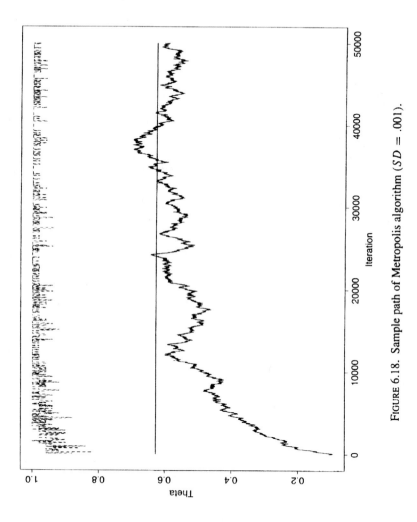

FIGURE 6.18. Sample path of Metropolis algorithm ($SD = .001$).

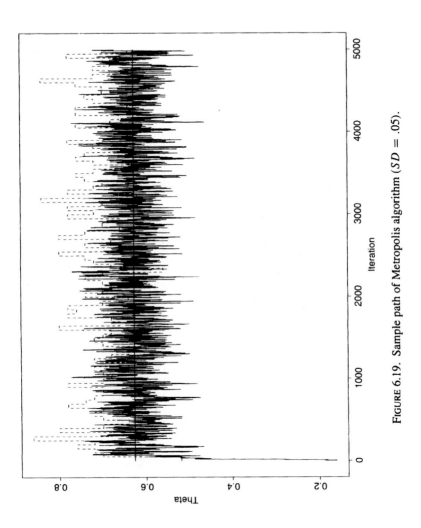

FIGURE 6.19. Sample path of Metropolis algorithm ($SD = .05$).

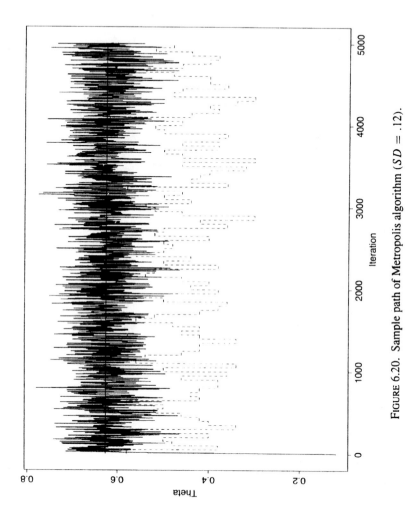

FIGURE 6.20. Sample path of Metropolis algorithm ($SD = .12$).

the posterior density *rather than quickly exploring the parameter space*. The two other chains quickly move into the region where the posterior mass is located.

Gelman et al. (1995) suggest that the optimal variance on the normal driver is $c^2 \Sigma$, where $c \approx 2.4/\sqrt{d}$ (d is the dimension of the parameter vector) and Σ is the variance-covariance matrix (i.e. based on the curvature of the posterior at the mode). In the present case $\Sigma = 0.05^2$ and $c = 2.4$, implying that the optimal standard deviation for the normal driver is 0.12. Gelman et al. (1995) also suggest that the optimal jumping rule has an acceptance rate of about 0.44 for one-dimensional problems, decreasing to 0.23 for problems where the dimension of the parameter vector exceeds 5. For the present example, the driver standard deviation of 0.12 does (approximately) yield the optimal rate. Exercise 2 examines other symmetric and asymmetric drivers. Convergence assessment techniques described in Section 6.3, such as Gelfand (1992), Gelman and Rubin (1992), or Geweke (1992), can be used in this case.

6.5.3. Metropolis Subchains

Müller (1993) suggests embedding a Metropolis random-walk subchain in the Gibbs sampler in the situation where it is difficult to sample directly from a conditional distribution, e.g. $p(\theta_1 \mid \theta_2, \ldots, \theta_d, Y)$. In particular, given the values from iteration i, $\theta_2^{(i)}, \theta_3^{(i)}, \ldots, \theta_d^{(i)}$, Müller (1993) lets $\pi(\rho) = p(\rho \mid \theta_2^{(i)}, \theta_3^{(i)}, \ldots, \theta_d^{(i)}, Y)$. The quantity $\theta_1^{(i+1)}$ is then the Tth value in the Metropolis chain:

a. Generate y^* from the distribution $f(y - x) = f(x - y)$, where x is the current state of the chain.
b. With probability $\alpha(x, y^*) = \min(1, \pi(y^*)/\pi(x))$ accept y^* as the new state; otherwise, the chain remains at x.

Müller (1993) suggests that the chain begin at $\theta_1^{(i)}$ and that $T = 1$; i.e. one pass through steps (a) and (b) is sufficient.

This algorithm, in contrast to rejection/acceptance algorithms, does not require the construction of a function that dominates $\pi(\rho)$ everywhere. However, it does require specification of the driving function $f(y - x)$. Like the griddy Gibbs sampler of Section 6.4, this algorithm relies on repeated evaluation of the conditional $p(\rho \mid \theta_2, \ldots, \theta_d, Y)$, modulo the normalizing constant. Tierney (1994) suggests using the grid-based approximation of $\pi(\rho)$ described in Section 6.4 to drive the Metropolis subchain. Gelman (1992) suggests other ways to construct these within-Gibbs subchains.

6.5.4. The Gibbs Sampler—Metropolis Connection

We now demonstrate that the Gibbs sampler is a generalized Metropolis algorithm where the probability of accepting the candidate value is always unity, following Gelman (1992). In other words, we address why in the Gibbs sampler we always accept, for example, θ_1^{t+1}, which is drawn from $p(\theta_1 \mid \theta_2^t, \ldots \theta_d^t, Y)$.

Since the Gibbs sampler is restricted to move along the coordinate axes, consider movement along the first coordinate axis. Denote the current point in the Gibbs sampler chain as $\theta = (\theta_1^t, \theta_2^t, \ldots \theta_d^t)$ and the candidate value as $\theta' = (\theta_1^{t+1}, \theta_2^t, \ldots \theta_d^t)$. Since the full conditionals are the drivers of the Gibbs sampler, the probability of going from θ to θ' is $p(\theta_1^{t+1} \mid \theta_2^t, \ldots \theta_d^t, Y)$. Similarly, the probability of going from θ' to θ is $p(\theta_1^t \mid \theta_2^t, \ldots \theta_d^t, Y)$. Hence, $\alpha(\theta, \theta')$ is equal to

$$
\frac{p(\theta_1^{t+1}, \theta_2^t, \ldots \theta_d^t \mid Y)}{p(\theta_1^t, \theta_2^t, \ldots \theta_d^t \mid Y)} \frac{p(\theta_1^t \mid \theta_2^t, \ldots \theta_d^t, Y)}{p(\theta_1^{t+1} \mid \theta_2^t, \ldots \theta_d^t, Y)}
$$

$$
= \frac{p(\theta_1^{t+1}, \theta_2^t, \ldots \theta_d^t, Y)}{p(\theta_1^t, \theta_2^t, \ldots \theta_d^t, Y)} \frac{p(\theta_1^t, \theta_2^t, \ldots \theta_d^t, Y)}{p(\theta_1^{t+1}, \theta_2^t, \ldots \theta_d^t \cdot, Y)} \frac{p(\theta_2^t, \ldots \theta_d^t, Y)}{p(\theta_2^t, \ldots \theta_d^t, Y)}
$$

$$
= 1.
$$

6.6 Conditional Inference via the Gibbs Sampler

6.6.1. Introduction

Several approaches are available for inference in the presence of nuisance parameters. This book has focused on the Bayesian/likelihoodist approach, where the nuisance parameters are eliminated by integration and inference is based on a marginal likelihood or marginal posterior distribution.

A Frequentist approach bases inference on the conditional distribution of the sufficient statistics corresponding to the parameters of interest, given the sufficient statistics corresponding to the remaining parameters fixed at their observed values. This device yields a distribution that is independent of the nuisance parameters in the exponential family case. As noted in Cox and Snell (1989), this is the standard Neyman–Pearson approach to the elimination of nuisance parameters. Andersen (1980) surveys the use of conditional likelihood functions to facilitate inference. The program of considering the exact permutation distribution of the conditional sufficient statistics in the case of logistic regression has been suggested by Cox (1971) and implemented in a computationally feasible manner by Bayer and Cox (1979), Hirji, Mehta and Patel (1987) and Tritchler (1984). In the contingency-table case, Patefield (1981) presents an algorithm for generating $r \times c$ contingency tables, conditional on row and column totals of the observed tables, to allow for a Monte Carlo conditional test of the model of independence. Mehta and Patel (1983, 1986a, b) present sophisticated network algorithms to perform exact calculations. In typical applications, however, hypothesis testing is generally performed by comparing the likelihood ratio, Wald or score statistic to the appropriate asymptotic distribution under the null; usually, a chi-square approximation is used. Confidence regions are typically constructed by inverting these approximate tests. It is well known that for small, highly stratified or unbalanced data sets, these asymptotic methods may be misleading.

Kolassa and Tanner (1994) propose a Gibbs–Skovgaard algorithm for the construction of a Markov chain whose equilibrium distribution $\tilde{F}$ approximates the distribution F of the sufficient statistics of interest, conditional on the sufficient statistics associated with the nuisance parameters. They approximate the univariate conditional distribution for a sufficient statistic corresponding to a canonical parameter of interest, conditional on all other sufficient statistics, using a saddlepoint approximation derived by Skovgaard (1987). The Gibbs sampler is then used to sample from this Markov chain with equilibrium distribution $\tilde{F}$. To generate a single chain realization, one draws an observation from the (Skovgaard) approximate distribution of each sufficient statistic conditional on all other sufficient statistics. Following the Gibbs scheme, one systematically scans all sufficient statistics of interest. Under certain regularity conditions, the resulting chain is ergodic and the equilibrium distribution is an approximation to the joint distribution of interest.

Their aim is to sample accurately from the multivariate conditional distribution of sufficient statistics corresponding to the null hypothesis. No considerations of power arise in their work; rather, Kolassa and Tanner (1994) present a method of calculating p values that are closer to those arising from the exact sampling distribution of the test statistic chosen than those arising from asymptotic methods.

Before proceeding, other uses of Markov chain methods in the Frequentist context are mentioned. Gelfand and Carlin (1993) and Geyer and Thompson (1992) illustrate how these methods can be used to locate the modes of the likelihood (or posterior). Besag and Clifford (1989) discuss the application of Markov chain methods to the problem of significance testing. *Assuming* one can devise an ergodic Markov chain whose equilibrium distribution is the same as the distribution of the sufficient statistic and whose transition probabilities are easy to sample from, Besag and Clifford (1989, 1991) suggest methods for computing p-values.

6.6.2. Skovgaard's Approximation

Skovgaard (1987) uses saddlepoint methods to derive approximate tail probabilities for conditional distributions. Let

$$K_X(\beta) = \log[E(\exp(\beta^T X))]$$

be the cumulant generating function of X, a sum of n independent random vectors. Denote by $\hat{\beta}$ the multivariate saddlepoint, which is determined by the saddlepoint equation:

$$K'_X(\hat{\beta}) = x,$$

where K'_X is the vector of derivatives of K_X and x is the observed value of X. Let $\hat{\gamma}$ be a vector of length d such that

$$K_{X_j}(\hat{\gamma}) = x_j \quad \text{for } j = 2, \ldots, d,$$
$$\hat{\gamma}_1 = 0,$$

where K_{X_j} denotes the derivative of K_X with respect to component j of its argument. Skovgaard (1987) approximates the conditional cumulative distribution function as

$$\Phi(\sqrt{n}\hat{w}) - \phi(\sqrt{n}\hat{w}) \left(\frac{\sqrt{|K''_{X^{-1}}(\hat{\gamma})|}}{\hat{\beta}_1\sqrt{n|K''_X(\hat{\beta})|}} - \frac{1}{\sqrt{n}\hat{w}} \right), \qquad (6.6.1)$$

where $\hat{\beta}_1$ is the first component of $\hat{\beta}$,

$$\hat{w} = \mathrm{sgn}(\hat{\beta}_1)\sqrt{2[\hat{\beta}^T x - K_X(\hat{\beta})] - 2[\hat{\gamma}^T x - K_X(\hat{\gamma})]},$$

$K''_{X^{-1}}$ is the $(d-1) \times (d-1)$ submatrix of the matrix of second derivatives of K_X, corresponding to all components of β and X except the first, and Φ and ϕ are the normal distribution function and density, respectively.

Also of interest are inversion techniques for lattice distributions. Skovgaard (1987) derives a counterpart of (6.6.1) in the lattice case:

$$\Phi(\sqrt{n}\hat{\omega}) - \phi(\sqrt{n}\hat{\omega}) \left(\frac{\sqrt{|K''_{X^{-1}}(\hat{\gamma})|}}{2\sinh(\frac{1}{2}\hat{\beta}_1)\sqrt{n|K''_X(\hat{\beta})|}} - \frac{1}{\sqrt{n}\hat{\omega}} \right) \qquad (6.6.2)$$

and x_1 is corrected for continuity when calculating $\hat{\beta}$. That is, if possible values for X_1 are Δ units apart, $\hat{\beta}$ solves $K'_X(\hat{\beta}) = \tilde{x}$, where $\tilde{x}_j = x_j$ if $j \neq 1$ and $\tilde{x}_1 = x_1 - \frac{1}{2}\Delta$. Skovgaard illustrates this formula in the context of the hypergeometric distribution, which has applications to testing independence in 2×2 tables.

6.6.3. Two-Way Contingency Tables

Inference on a subset of parameters in a regression model in the presence of other parameters is often difficult, especially in the case of nonlinear regression models where the device of orthogonality is not available. Cox and Snell (1989) suggest conditioning on the sufficient statistics corresponding to regression parameters not of immediate interest. The situation in this section exhibits approximate tests of the hypothesis of independence in contingency tables. This situation represents a special case of Poisson regression, in which an $r \times c$ table of independent Poisson random variables is observed. One assumes a log-linear model for the means with row and column effects and row $\times$ column interactions. The test of independence is equivalent to testing whether these interactions are simultaneously zero. Row and column total means are nuisance parameters; hence, we condition on the observed row and column totals, to test the hypothesis of independence. Conditioning on row and column totals leaves random $(r-1) \times (c-1)$ sufficient statistics. These sufficient statistics not conditioned on may be chosen to be table entries in all but the last row and column. Denote these as $T_1, \ldots, T_{(r-1)(c-1)}$. The idea in Kolassa and Tanner (1994) is to generate a Markov chain whose equilibrium distribution is approximately the joint distribution of $T_1, \ldots, T_{(r-1)(c-1)}$, conditional on the row and column totals. They use the Gibbs sampler to reduce this to the problem of sampling from a distribution approximating that of T_i conditional on T_j $\forall j \neq i$ and the row and column totals.

Kolassa and Tanner (1994) apply the Gibbs–Skovgaard algorithm to test the model of independence in an $r \times c$ contingency table $\mathbf{X} = (X_{ij})$, where X_{ij} are independent Poisson random variables with means $\alpha + \beta_i + \gamma_j + \delta_{ij}$. As presented here, this model is overparameterized. Constrain these parameters so that $\beta_r = \gamma_c = 0$ and $\delta_{ij} = 0$ if $i = r$ and $j = c$. The rc sufficient statistics for the saturated model can be chosen to be the raw table entries $\mathbf{X}$. However, an alternative selection of sufficient statistics will make implementing the Gibbs–Skovgaard algorithm more convenient. This algorithm requires sampling conditional on the canonical sufficient statistics associated with main, row and column effects α, β_i and γ_j. Hence, a collection $\mathbf{T}$ of sufficient statistics must be found to include these canonical sufficient statistics. These sufficient statistics are the grand total, the $r - 1$ row totals for all but the last row, and the $c - 1$ column totals for all but the last column. Canonical sufficient statistics for the parameters δ_{ij} are the $(r - 1) \times (c - 1)$ table entries in all but the last row and column. Let $\mathbf{T}$ be this vector of sufficient statistics.

The Gibbs sampler is used to sample from the Markov chain whose equilibrium distribution is approximately the joint distribution of these $(r - 1) \times (c - 1)$ table entries $(T_1, \ldots, T_{(r-1)(c-1)})$ conditional on the other $r + c - 1$ sufficient statistics. One could proceed by drawing random observations uniform on $(0, 1)$ and inverting the conditional cumulative distribution function approximation (6.6.2) to get a random observation from a distribution approximating the true conditional distribution. In this case, (6.6.2) would be calculated with $K_\mathbf{T}$ in place of K_X. Kolassa and Tanner (1994), however, show how this may be simplified. First, they show how the desired conditional distribution may be generated by a derived 2×2 table rather than by the original larger $r \times c$ table. Then they present an alternative to the approximation (6.6.2) not requiring the joint cumulant generating function of a table observation and the marginals but only that of raw table entries.

Kolassa and Tanner (1994) prove a theorem that allows one to reduce the problem of sampling from the conditional distribution of an entry in this $r \times c$ table to the problem of sampling from a hypergeometric distribution arising from a 2×2 table. More formally, their theorem states that the distribution of one of these first $(r - 1) \times (c - 1)$ components of $\mathbf{T}$, the entry in row i and column j of the table $\mathbf{X}$, conditional on all other sufficient statistics, is the same as that if X_{ij} has a hypergeometric distribution generated by the derived 2×2 table

$$
\begin{array}{|cc|c}
\hline
X_{ij} & x_{ic} & x_{ij} + x_{ic} \\
x_{rj} & x_{rc} & x_{rj} + x_{rc} \\
\hline
x_{ij} + x_{rj} & x_{ic} + x_{rc}
\end{array}
\qquad (6.6.3)
$$

conditional on row and column totals for this reduced table.

The idea in Kolassa and Tanner (1994) is to sample approximately from this conditional distribution by drawing an observation from a uniform distribution on (0, 1) and inverting an approximation to the conditional cumulative distribution function arising in (6.6.3) at this random observation. They show that the Skovgaard (1987) approximation in this case can be reduced to

$$P(X_{ij} \le x_{ij} \mid R < C < N)$$

$$= \Phi(\hat{\omega}) - \phi(\hat{\omega}) \left\{ \left(\frac{x_{i.} x_{.j} x_{r.} x_{.c}}{x_{..} \tilde{x}_{ij} \tilde{x}_{ic} \tilde{x}_{rj} \tilde{x}_{rc}} \right)^{1/2} [2 \sinh(\rho)]^{-1} - \hat{\omega}^{-1} \right\},$$

where $\rho = \log[\tilde{x}_{ij} \tilde{x}_{rc}/(\tilde{x}_{rj} \tilde{x}_{ic})]$ is the log of the observed continuity-corrected odds ratio, $R = x_{ij} + x_{ic}$, $C = x_{ij} + x_{rj}$, $N = x_{ij} + x_{ic} + x_{rj} + x_{rc}$, $\tilde{x}_{ij} = x_{ij} - \frac{1}{2}$, $\tilde{x}_{rj} = x_{rj} + \frac{1}{2}$, $\tilde{x}_{ic} = x_{ic} + \frac{1}{2}$, $\tilde{x}_{rc} = x_{rc} - \frac{1}{2}$, and

$$\hat{\omega} = \text{sgn}(\rho) \left\{ 2 \left[\sum_{\substack{l=i,r \\ m=j,c}} \tilde{x}_{lm} \log(\tilde{x}_{lm}) - \sum_{l=i,r} x_{l.} \log(x_{l.}) \right. \right.$$

$$\left. \left. - \sum_{m=j,c} x_{.m} \log(x_{.m}) + x_{..} \log(x_{..}) \right] \right\}^{1/2}.$$

In this case, · denotes summation over the reduced table: $x_{l.} = x_{lj} + x_{lc}$, $x_{.m} = x_{im} + x_{rm}$, and $x_{..} = x_{ij} + x_{ic} + x_{rj} + x_{rc}$.

This equation provides an approximation to the distributions of each of the sufficient statistics representing individual table counts conditional on all other sufficient statistics. Random observations from a Markov chain are generated by cycling through the Gibbs sampling procedure and at each stage drawing a random sample from the cumulative distribution function given above. Cycling through the Gibbs sampling scheme many times will give a multivariate sample whose distribution is close to the equilibrium distribution, which in turn is an approximation to the joint distribution of the sufficient statistics representing table counts conditional on row and column totals. Kolassa and Tanner (1994) also consider the case of three-way tables and of logistic regression.

6.6.4. Example

This example is a 4 × 4 contingency table of responses to the questionnaire item "Sex is fun for me and my partner (a) never or occasionally, (b) fairly often, (c) very often, (d) almost always" for 91 married couples from the Tucson metropolitan area, as reported by Hout, Duncan and Sobel (1987) and cited by Agresti (1990) (see Table 6.4). These data were chosen to assess the quality of the Markov chain algorithm in a situation in which usual asymptotic approximations may be inappropriate. Again, the hypothesis of independence was tested by generating random tables using the Gibbs sampler and Skovgaard's approximation as described above. The statistic for the likelihood ratio test was calculated for each simulated table.

TABLE 6.4.   Sexual Response Data.

Husband's response	Wife's Response			
	Never or occasionally	Fairly often	Very often	Almost always
Never or occasionally	7	7	2	3
Fairly often	2	8	3	7
Very often	1	5	4	9
Almost always	2	8	9	14

Kolassa and Tanner (1994) simulated 50,000 independent Markov chains for 30 iterations. For each integer k between 1 and 30, they estimated the p-value after iteration k generated by the test statistic by calculating the test statistic for the observed table, and for each of the 50,000 tables represented by the state of the chain at time k. They report as the estimated p-value the proportion of the sample tables with a test-statistic value as high or higher than the observed value. Convergence was assessed by monitoring these p-values across iterations; in this example, note that after fewer than 10 iterations the estimated p-values become stable. These results were compared (see Table 6.5) with those obtained from the StatXact implementation of the Monte Carlo algorithm due to Patefield (1981), using 50,000 samples. The results are given in Table 6.5. The p-value based on the final 9000 iterations of one chain of length 10,000 was 0.109.

The Gibbs–Skovgaard algorithm represents an improvement over the asymptotic value in that the Gibbs–Skovgaard p value lies closer to the true p value than does the p value derived from the χ^2 approximation. It is not argued that the Gibbs–Skovgaard algorithm is the preferred algorithm for testing the model of independence in the $r \times c$ contingency table. The goal is to illustrate the accuracy of this general algorithm in a situation where the exact (up to Monte Carlo error) result is known. This work provides a *general algorithm* to handle small sample inference in the exponential family.

TABLE 6.5.   Results for Sexual Response Data.

Test method	Likelihood ratio
StatXact Monte Carlo p value	0.111
Gibbs–Skovgaard p value	0.107
Asymptotic p value	0.078

Exercises

1. Recall the genetic linkage model of Section 5.1.

   a. For the data $Y = (125, 18, 20, 34)$, implement the Gibbs sampler algorithm. Use a flat prior on θ. Plot θ^i versus iteration i. How long a chain (or chains) did you use? Use the Gelman and Rubin (1992) and Geweke (1992) methods to assess convergence. Did you toss out any initial values? Compute the posterior mean and posterior variance based on your chain. Plot the estimated observed posterior along with the normalized likelihood. Discuss the adequacy of the estimate.

   b. Repeat (a) for the data $Y = (14, 0, 1, 5)$.

   c. Run 20 chains with independent starting values. Compute the average of the θ's in each chain. Calculate the standard deviation of the 20 averages. Interpret this value. Compute the standard deviation of the θ's in each chain. Divide each SD by the square root of the number of iterations. Average these "standard errors" and compare this average to the standard deviation of the averages. Would you expect these numbers to be similar or different?

2. a. For the genetic linkage model applied to the data $Y = (125, 18, 20, 34)$, implement the Metropolis algorithm (use a flat prior on θ). Use one long chain and plot θ^i versus i. Try several driver functions: (1) uniform on $(0,1)$; (2) normal centered at the current point in the chain and standard deviation 0.01; (3) normal with same mean as in (2) and standard deviation 0.1; (4) normal with same mean as in (2) and standard deviation 0.5; (5) normal centered at 0.4 with standard deviation 0.1. Draw a histogram of the θ values and the true normalized likelihood in the same figure. Discuss the adequacy of the Metropolis estimate. How did you assess convergence? Did you toss out any initial values? Should you use the Metropolis or generalized Metropolis algorithm in part (5)?

   b. Repeat (a) for the data $Y = (14, 0, 1, 5)$.

   c. Compute both the posterior mean and standard deviation for both datasets. Compare to the results from the previous problem.

   d. For each of the drivers in part (a), run 20 chains with independent starting values. Compute the average of the θ's in each chain. Calculate the standard deviation of the 20 averages. Interpret this value. Compute the standard deviation of the θ's in each chain. Divide each SD by the square root of the number of iterations. Average these "standard errors" and compare this average to the standard deviation of the averages. Would you expect these numbers to be similar or different? Compare to the results of Exercise 1c.

3. a. Consider the 1-way variance components model

$$Y_{ij} = \theta_i + \varepsilon_{ij},$$

where Y_{ij} is the jth observation from the ith group, θ_i is the effect of the ith group and ε_{ij} is the error, where $i = 1, \ldots, K$ and $j = 1, \ldots, J$. It is assumed that the ε_{ij} are an iid sample from $N(0, \sigma_\varepsilon^2)$ and the θ_i are an independent iid sample from $N(\mu, \sigma_\theta^2)$. Under the prior specification

$p(\sigma_\varepsilon^2, \sigma_\theta^2, \mu) = p(\sigma_\varepsilon^2)p(\sigma_\theta^2)p(\mu)$, with $p(\sigma_\theta^2) = IG(a_1, b_1)$, $p(\sigma_\varepsilon^2) = IG(a_2, b_2)$ and $p(\mu) = N(\mu_0, \sigma_0^2)$, show that

i. $p(\mu \mid \theta, \sigma_\varepsilon^2, \sigma_\theta^2, Y) = N\left(\frac{\sigma_\theta^2 \mu_0 + \sigma_0^2 \Sigma \theta_i}{\sigma_\theta^2 + K\sigma_0^2}, \frac{\sigma_\theta^2 \sigma_0^2}{\sigma_\theta^2 + K\sigma_0^2} \right)$

ii. $p(\theta_i \mid \mu, \sigma_\varepsilon^2, \sigma_\theta^2, Y) = N\left(\frac{J\sigma_\theta^2}{J\sigma_\theta^2 + \sigma_\varepsilon^2} \bar{Y}_{i.} + \frac{\sigma_\varepsilon^2}{J\sigma_\theta^2 + \sigma_\varepsilon^2} \mu, \frac{\sigma_\theta^2 \sigma_\varepsilon^2}{J\sigma_\theta^2 + \sigma_\varepsilon^2} \right)$,

iii. $p(\sigma_\varepsilon^2 \mid \mu, \theta, \sigma_\theta^2, Y) = IG\left(a_2 + \frac{KJ}{2}, b_2 + \frac{1}{2} \sum_{i=1}^{K} \sum_{j=1}^{J} (Y_{ij} - \theta_i)^2 \right)$

iv. $p(\sigma_\theta^2 \mid \mu, \theta, \sigma_\varepsilon^2, Y) = IG\left(a_1 + \frac{K}{2}, b_1 + \frac{1}{2} \sum_{i=1}^{K} (\theta_i - \mu)^2 \right)$,

where $\bar{Y}_{i.} = \frac{1}{J} \sum_{j=1}^{J} Y_{ij}$ and $\theta = (\theta_1, \ldots, \theta_K)$.

b. Run the Gibbs sampler for the data below. Use one chain of length 75,000. Take $p(\mu) = N(0, 10^{12})$, $p(\sigma_\varepsilon^2) = IG(0, 0)$ and $p(\sigma_\theta^2) = IG(1, 1)$. For each θ_i, for σ_ε, and for σ_θ, plot the simulated value at iteration j versus j. Summarize each posterior marginal.
c. Repeat (b) using the prior specification $p(\mu) = N(0, 10^{12})$, $p(\sigma_\varepsilon^2) = IG(0, 0)$ and $p(\sigma_\theta^2) = IG(0, 0)$. Does this specification violate the Hobert–Casella conditions? Describe what happens to the Gibbs sampler chain in this case.
d. Repeat (b) with $p(\sigma_\varepsilon^2) = constant$ and $p(\sigma_\mu^2) = constant$.

Data Generated from a Table of Random Normal Deviates (Box and Tiao, 1973).

Batch	Observations				
1	7.298	3.846	2.434	9.566	7.990
2	5.220	6.556	0.608	11.788	-0.892
3	0.110	10.386	13.434	5.510	8.166
4	2.212	4.852	7.092	9.288	4.980
5	0.282	9.014	4.458	9.446	7.198
6	1.722	4.782	8.106	0.758	3.758

4. Recall the Darwin data of Exercise 2 in Chapter 2.
   a. Suppose that the data follow the t distribution with mean μ, variance σ^2 (unknown), on 4 degrees of freedom. Under the noninformative prior, use the Metropolis algorithm to obtain the posterior marginal of μ and of σ. Use the Geweke (1992) technique to assess convergence.
   b. Repeat (a) using multiple chains started from independent starting points. Use the Gelman and Rubin (1992) approach to assess convergence.
   c. Repeat (a) assuming the degrees of freedom on the t distribution are unknown.

5. For the radiotherapy data of Section 2.3.2, analyze the data using the logistic regression model (use the Metropolis algorithm). Obtain the posterior marginal for α and for β under a flat prior for α and for β. Compute the posterior marginal mean for α and for β. Compute the posterior marginal standard deviation for α and for β. How do these results compare to the normal approximation results of Section 2.3.2? Obtain the posterior distribution of $-\frac{\alpha}{\beta}$. What is the interpretation of this quantity?

6. a. Derive the full conditionals for the item response model discussed in Section 6.2.5.

   b. Thirty-nine students took a quiz in an introductory statistics course consisting of six questions. Below are the results. A correct answer was scored as a 1 and an incorrect answer was scored as a 0. Using the Gibbs sampler approach, apply the item response model to these data. How did you assess convergence? Obtain the posterior marginal for α_i and β_i for $i = 1, \ldots, 6$. What is the posterior mean and standard deviation for each of these parameters? Which questions are hard? Which are easy? Which are good discriminators? Rank the students according to the θ_i marginals.

Student	Q1	Q2	Q3	Q4	Q5	Q6
1	1	0	1	0	0	1
2	1	1	1	0	1	1
3	1	0	0	0	1	0
4	0	1	0	0	0	0
5	1	1	1	0	0	1
6	0	0	1	0	1	1
7	1	1	1	1	0	0
8	1	1	1	1	1	1
9	0	1	0	0	1	1
10	0	1	1	0	1	1
11	0	1	0	0	0	0
12	0	1	0	1	0	1
13	0	1	0	0	0	1
14	1	0	1	0	0	1
15	1	0	1	0	0	0
16	0	1	0	0	0	0
17	0	1	1	0	1	1
18	0	0	1	1	0	1
19	1	1	1	0	1	1
20	0	0	1	0	0	0

Student	Q1	Q2	Q3	Q4	Q5	Q6
21	1	1	0	0	0	1
22	0	1	0	0	0	0
23	0	1	1	1	0	0
24	1	1	1	1	1	1
25	1	0	0	0	0	0
26	1	0	0	0	0	0
27	0	1	0	0	0	1
28	1	1	1	1	1	0
29	1	0	0	1	1	0
30	1	1	0	0	0	0
31	1	0	0	0	0	0
32	0	0	1	1	0	0
33	0	0	0	1	1	0
34	0	0	0	0	0	1
35	1	0	0	0	0	0
36	0	1	0	0	1	1
37	1	0	0	1	1	1
38	1	1	1	1	1	1
39	1	0	0	0	1	1

7. The data below are the results from a mutagenicity assay on *Salmonella* in which three plates were processed at each dose i of quinoline (Breslow, 1984). Let x_i be the dose on the plates $i1$, $i2$ and $i3$ and let y_{ij} follow a Poisson distribution with mean μ_{ij}, where

$$\log(\mu_{ij}) = \alpha + \beta \log(x_i + 10) + \gamma x_i + \lambda_{ij},$$

and λ_{ij} is normally distributed with mean 0 and (unknown) variance σ^2.

a. Under the prior $p(\alpha, \beta, \gamma, \sigma^2) \propto p(\alpha)p(\beta)p(\gamma)p(\sigma^2) = p(\sigma^2) = IG(a, b)$, where a and b are known, derive the full conditionals for use in the Gibbs sampler.

b. Run the Gibbs sampler for this problem (take $a = b = 0.001$). How did you assess convergence? How many chains did you use? Under the prior specification, is the posterior proper? Obtain the posterior marginals for α, β, γ, and σ^2 via a Rao-Blackwell estimate. Compute the posterior marginal means and standard deviations for α, β, γ, and σ^2.

c. Repeat (b) for $a = b = 0$.

d. Modify your random effects Poisson model allowing for overdispersion, where λ_{ij} now follows the t distribution on 4 df with mean zero and unknown variance σ^2. Repeat (a) and (b).

Dose of Quinoline (μg per plate)					
0	10	33	100	333	1000
15	16	16	27	33	20
21	18	26	41	38	27
29	21	33	60	41	42

8. Redo Exercise 8 of Chapter 3 using the Metropolis algorithm to sample from $p(\alpha, \beta \mid Y)$. Compare these results to those obtained via importance sampling.

References

Agresti, A. (1990). *Categorical Data Analysis*. New York: Wiley Interscience.

Aitkin, M. (1981). A note on regression analysis of censored data. *Technometrics* **23**, 161–163.

Albert, J.H. (1992). Bayesian estimation of normal ogive item response curves using Gibbs sampling. *Journal of Educational Statistics* **17**, 251–269.

Albert, J.H. and Chib, S. (1993). Bayesian analysis of binary and polychotomous response data. *Journal of the American Statistical Association* **88**, 669–679.

Albert, J.H. and Chib, S. (1995). Bayesian residual analysis for binary response regression models. *Biometrika* **82**, 747–759.

Amit, Y. (1991). On rates of convergence of stochastic relaxation for Gaussian and non-Gaussian distributions. *Journal of Multivariate Analysis* **38**, 82–99.

Andersen, E.B. (1980). *Discrete Statistical Models*. Amsterdam: North-Holland.

Applegate, D., Kannan, R., and Polson, N. (1990). Random polynomial time algorithms for sampling from joint distributions. Technical report, School of Computer Science, Carnegie Mellon University.

Barker (1965). Monte Carlo calculations of radial distribution functions for a proton-electron plasma. *Australian Journal of Physics* **18**, 119–133.

Bates, D.M. and Watts, D.G. (1988). *Nonlinear Regression Analysis and its Applications*. New York: Wiley.

Bayer, L. and Cox, C. (1979). Algorithm AS142. *Applied Statistics* **28**, 319–324.

Behrens, W.A. (1929). Ein betrag zur fehlenberechnung bei wenigen beobachtungen. *Landwirtschaftliche Jahrbücher* **68**, 807–837.

Berger, J.O. and Bernardo, J.M. (1992). On the development of reference priors (with discussion). In: *Bayesian Statistics 4*. J.M. Bernardo, J.O. Berger, A.P. Dawid, and A.F.M. Smiths (eds), Oxford: Oxford University Press.

Besag, J. (1974). Spatial interaction and the statistical analysis of lattice systems. *Journal of the Royal Statistical Society B* **36**, 192–236.

Besag, J. and Clifford, P. (1989). Generalized Monte Carlo significance tests. *Biometrika* **76**, 633–642.

Besag, J. and Clifford, P. (1991). Sequential Monte Carlo p-values. *Biometrika* **78**, 301–304.

Besag, J. and Green, P.J. (1993). Spatial statistics and Bayesian computation. *Journal of the Royal Statistical Society B* **55**, 25–38.

Besag, J., Green, P., Higdon, D., and Mengersen, K. (1995). Bayesian computation and stochastic systems (with discussion). *Statistical Science* **10**, 3–66.

Bickel, P.J. and Doksum, K.A. (1977). *Mathematical Statistics*. Oakland: Holden-Day.

Bishop, Y.M.M., Fienberg, S., and Holland, P. (1975). *Discrete Multivariate Analysis*. Cambridge: MIT Press.

Box, G.E.P. and Tiao, G. (1973). *Bayesian Inference in Statistical Analysis*. Reading: Addison-Wesley.

Breslow, N.E. (1984). Extra Poisson variation in log-linear models. *Applied Statistics* **33**, 38–44.

Brown, C.H. (1990). Asymptotic properties of estimators with nonignorable missing data. Technical Report, Department of Biostatistics, Johns Hopkins.

Buonaccorsi, J.P. and Gatsonis, C.A. (1988). Bayesian inference for ratios of coefficients in a linear model. *Biometrics* **44**, 87–102.

Carlin, B.P. and Gelfand, A.E. (1991). An iterative Monte Carlo method for nonconjugate Bayesian analysis. *Statistics and Computing* **1**, 119–128.

Carlin, B.P., Gelfand, A.E., and Smith, A.F.M. (1992). Hierarchical Bayesian analysis of change point problems. *Journal of the Royal Statistical Society C* **41**, 389–405.

Carlin, B.P. and Polson, N.G. (1991). Inference for nonconjugate Bayesian models using the Gibbs sampler. *Canadian Journal of Statistics* **19**, 399–405.

Carlin, B.P., Polson, N.G., and Stoffer, D.S. (1992). A Monte Carlo approach to nonnormal and nonlinear state space modelling. *Journal of the American Statistical Association* **87**, 493–500.

Carter, C.K. and Kohn, R. (1994). On Gibbs sampling for state space models. *Biometrika* **81**, 541–553.

Casella, G. and George, E.I. (1992). Explaining the Gibbs sampler. *American Statistician* **46**, 167–174.

Chan, K.S. (1993). Asymptotic behavior of the Gibbs sampler. *Journal of the American Statistical Association* **88**, 320–326.

Chan, K.S. and Ledholter, J. (1995). Monte Carlo EM estimation for time series models involving counts. *Journal of the American Statistical Association* **90**, 242–252.

Chen, M.H. (1994). Importance-weighted marginal Bayesian posterior density estimation. *Journal of the American Statistical Association* **89**, 818–824.

Chen, M.H. and Schmeiser, B. (1993). Performance of the Gibbs, hit-and-run and Metropolis samplers. *Journal of the Computational and Graphical Statistics* **2**, 251–272.

Chen, R. and Liu, J. (1996). Predictive updating methods with application to Bayesian classification. *Journal of the Royal Statistical Society B* **58**, 397–415.

Chib, S. (1992). Bayes inference in the Tobit censored regression model. *Journal of Econometrics* **51**, 79–99.

Chib, S. (1995). Marginal likelihood from the Gibbs output. *Journal of the American Statistical Association* **90**, 1313–1321.

Chib, S. and Greenberg, E. (1995). Understanding the Metropolis-Hastings algorithm. *American Statistician* **49**, 327–335.

Clarke, B. and Wasserman, L. (1993). Noninformative priors and nuisance parameters. *Journal of the American Statistical Association* **88**, 1427–1432.

Clayton, D.G. (1991). A Monte Carlo method for Bayesian inference in frailty models. *Biometrics* **47**, 467–485.

Cochran, W.G. (1977). *Sampling Techniques*. New York: Wiley.

Cook, D. and Weisberg, S. (1982). *Residuals and Influence in Regression*. London: Chapman and Hall.

Cox, D.R. (1971). *Analysis of Binary Data*. London: Chapman and Hall.

Cox, D.R. and Hinkley, D.V. (1974). *Theoretical Statistics*. London: Chapman and Hall.

Cox, D.R. and Snell, E.J. (1989). *Analysis of Binary Data*. London: Chapman and Hall.

Cowles, M.K. and Carlin, B.P. (1996). Markov chain Monte Carlo convergence diagnostics: A comparative review. *Journal of the American Statistical Association*, (to appear).

Cowles, M.K., Carlin, B.P., and Connett, J.E. (1996). Bayesian Tobit modeling of longitudinal ordinal clinical trial data with nonignorable missingness. *Journal of the American Statistical Association*, (to appear).

Cui, L., Tanner, M.A., Sinha, D., and Hall, W.J. (1992). Monitoring convergence of the Gibbs sampler: Further experience with the Gibbs stopper. *Statistical Science* **7**, 483–486.

Datta, G.S. and Ghosh, J.K. (1995). On priors providing frequentist validity for Bayesian inference. *Biometrika* **82**, 37–45.

Davis, P. and Rabinowitz, P. (1984). *Methods of Numerical Integration*. New York: Academic Press.

Dellaportas, P. (1995). Random variate transformations in the Gibbs sampler. *Statistics and Computing* **5**, 133–140.

Dellaportas, P. and Smith, A.F.M. (1993). Bayesian inference for generalized linear and proportional hazards models via Gibbs sampling. *Applied Statistics* **42**, 443–460.

Dempster, A.P., Laird, N., and Rubin, D.B. (1977). Maximum likelihood from incomplete data via the EM algorithm. *Journal of the Royal Statistical Society B* **39**, 1–38.

Diaconis, P. and Ylvisaker, D. (1979). Conjugate priors for exponential families. *Annals of Statistics* **7**, 269–281.

Diebolt, J. and Robert, C. (1994). Estimation of finite mixture distributions. *Journal of the Royal Statistical Society B* **56**, 363–375.

Efron, B. and Hinkley, D.V. (1978). Assessing the accuracy of the maximum likelihood estimator: Observed versus expected Fisher information. *Biometrika* **65**, 457–482.

Fisher, R.A. (1935). The fiducial argument in statistical inference. *Annals of Eugenics* **6**, 391–398.

Frigessi, A., di Stefano, P., Hwang, C.R., and Sheu, S.J. (1993). Convergence rates of the Gibbs sampler, the Metropolis algorithm and other single-site updating dynamics. *Journal of the Royal Statistical Society B* **55**, 205–220.

Gelfand, A.E. (1992). Discussion to the papers of Gelman and Rubin and of Geyer. *Statistical Science* **4**, 486–487.

Gelfand, A.E. and Carlin, B.P. (1993). Maximum likelihood estimation for constrained or missing data model. *Canadian Journal of Statistics* **21**, 303–311.

Gelfand, A.E., Hills, S.E., Racine-Poon, A., and Smith, A.F.M. (1990). Illustration of Bayesian inference in normal data models using Gibbs sampling. *Journal of the American Statistical Association* **85**, 972–985.

Gelfand, A.E. and Sahu, S.K. (1994). On Markov chain Monte Carlo acceleration. *Journal of the Computational and Graphical Statistics* **3**, 261–276.

Gelfand, A.E. and Smith, A.F.M. (1990). Sampling-based approaches to calculating marginal densities. *Journal of the American Statistical Association* **85**, 398–409.

Gelfand, A.E. and Smith, A.F.M. (1991). Gibbs sampling for marginal posterior expectations. *Communications in Statistics A* **20**, 1747–1766.

Gelfand, A.E., Smith, A.F.M., and Lee, T.M. (1992). Bayesian analysis of constrained parameter and truncated data problems. *Journal of the American Statistical Association* **87**, 523–532.

Gelman, A. (1992). Iterative and non-iterative simulation algorithms. In: *Computing Science and Statistics: Proceedings of the 24th Symposium on the Interface*. Fairfax: Interface Foundation of North America.

Gelman, A. and Rubin, D.B. (1992). Inference from iterative simulation using multiple sequences. *Statistical Science*. **7**, 457–472.

Gelman, A., Carlin, J.B., Stern, H.S., and Rubin, D.B. (1995). *Bayesian Data Analysis*. London: Chapman and Hall.

Geman, S. and Geman, D. (1984). Stochastic relaxation, Gibbs distributions and the Bayesian restoration of images. *IEEE Transactions on Pattern Analysis and Machine Intelligence* **6**, 721–741.

George, E.I. and Robert, C.P. (1992). Capture-recapture estimation via Gibbs sampling. *Biometrika* **79**, 677–683.

Geweke, J. (1989). Bayesian inference in econometric models using Monte Carlo integration. *Econometrica* **57**, 1317–1339.

Geweke, J. (1992). Evaluating the accuracy of sampling-based approaches to the calculation of posterior moments. In: *Bayesian Statistics 4*. J.M. Bernardo, J.O. Berger, A.P. Dawid and A.F.M. Smiths (eds), Oxford: Oxford University Press.

Geyer, C. (1992). Practical Markov chain Monte Carlo. *Statistical Science* **7**, 473–482.

Geyer, C.J. and Thompson, E.A. (1992). Constrained Monte Carlo maximum likelihood for dependent data (with discussion). *Journal of the Royal Statistical Society Series B* **54**, 657–699.

Gilks, W.R. and Wild, P. (1992). Adaptive rejection sampling for Gibbs sampling. *Journal of the Royal Statistical Society C* **41**, 337–348.

Gilks, W.R., Clayton, D.G., Spiegelhalter, D.J., Best, N.G., McNeil, A.J., Sharples, L.D., and Kirby, A. J. (1993). Modelling complexity. *Journal of the Royal Statistical Society B* **55**, 39–52.

Glasser, M. (1965). Regression analysis with dependent variable censored. *Biometrics* **21**, 300-307.

Glynn, R.J., Laird, N.M., and Rubin, D.B. (1986). Mixing modelling versus selection modelling with nonignorable nonresponse. In: H. Wainer (ed.). *Drawing Inferences from Self-Selected Samples*. New York: Springer-Verlag.

Glynn, R.J., Laird, N.M., and Rubin, D.B. (1993). Multiple imputation in mixture models for nonignorable nonresponse with follow-ups. *Journal of the American Statistical Association* **88**, 984–993.

Goldman, A.I., Carlin, B.P., Crane, L.R., Launer, C., Korvick, J.A., Deyton, L., and Abrams, D.I. (1996). Response of CD4+ and clinical consequences to treatment using ddI or ddC in patients with advanced HIV infection. *Journal of Acquired Immune Deficiency Syndromes and Human Retrovirology*, (to appear).

Goodman, L.A. (1974). Exploratory latent structure analysis using both identifiable and unidentifiable models. *Biometrika* **61**, 215–231.

Greenlees, J.S., Reece, W.S., and Zieschang, K.Y. (1982). Imputation of missing values. *Journal of the American Statistical Association* **77**, 251–261.

Guo, S.W. and Thompson, E.A. (1991). Monte Carlo estimation of variance component models. *IMA Journal of Mathematics Applied in Medicine and Biology* **8**, 171–189.

Guo, S.W. and Thompson, E.A. (1994). Monte Carlo estimation of mixed models for large complex pedigrees. *Biometrics* **50**, 417–432.

Haberman, S.J. (1978). *Analysis of Qualitative Data*, Vol. 1, New York: Academic Press.

Haberman, S.J. (1979). *Analysis of Qualitative Data*, Vol. 2, New York: Academic Press.

Haldane, J.B.S. (1948). The precision of observed values of small frequencies. *Biometrika* **35**, 297–303.

Hammersley, J.M. and Handscomb, D.C. (1964). *Monte Carlo Methods*. London: Chapman and Hall.

Hannan, E.J. (1977). *Time Series Analysis*. London: Chapman and Hall.

Hastings, W.K. (1970). Monte Carlo sampling methods using Markov chains and their applications. *Biometrika* **57**, 97–109.

Hirji, K.F., Mehta, C.R., and Patel, N.R. (1987). Computing distributions for exact logistic regression. *Journal of the American Statistical Association* **82**, 1110–1117.

Hobert, J.P. and Casella, G. (1993). Gibbs sampling with improper prior distributions. Technical Report, Biometrics Unit, Cornell University.

Hout, M., Duncan, O.D., and Sobel, M.E. (1987). Association and heterogeneity: structural models of similarities and differences. *Sociological Methodology* **17**, 145–184.

Jamshidian, M. and Jennrich, R. (1993). Conjugate gradient acceleration of the EM algorithm. *Journal of the American Statistical Association* **88**, 221–228.

Jeffreys, H.S. (1961). *Theory of Probability*. Oxford: Oxford University Press.

Kass, R.E. and Raftery, A. (1995). Bayes factors. *Journal of the American Statistical Association* **90**, 773–795.

Kass, R.E., Tierney, L., and Kadane, J.B. (1989). Approximate methods for assessing influence and sensitivity in Bayesian analysis. *Biometrika* **76**, 663–674.

Kass, R. E. and Wasserman, L. (1993). Formal rules for selecting prior distributions: A review. Technical Report. Carnegie Mellon University.

Kolassa, J.E. and Tanner, M.A. (1994). Approximate conditional inference via the Gibbs sampler. *Journal of the American Statistical Association* **89**, 697–702.

Kong, A., Liu, J., and Wong, W.H. (1994). Sequential imputations and Bayesian missing data problems. *Journal of the American Statistical Association* **89**, 278–288.

Kuo, L. and Yang, T. (1995). Bayesian computation of software reliability. *Journal of the Computational and Graphical Statistics* **4**, 65–82.

Lawless, J.F. (1982). *Statistical Models and Methods for Lifetime Data*. New York: Wiley.

Lange, K. (1995a). A gradient algorithm locally equivalent to the EM algorithm. *Journal of the Royal Statistical Society B* **57**, 425–437.

Lange, K. (1995b). A quasi-Newton acceleration of the EM algorithm. *Statistica Sinica* **5**, 1–18.

Lawrence, C.E., Altschul, S.F., Boguski, M.S. Liu, J.S., Neuwald, A., and Wooton, J. (1993). Detecting subtle sequence signals: A Gibbs sampling strategy for multiple alignment. *Science* **262**, 208–214.

Lee, P.M. (1992). *Bayesian Statistics*. New York: Wiley.

Leonard, T., Hsu, J.S.J., and Tsui, K.W. (1989). Bayesian marginal inference. *Journal of the American Statistical Association* **84**, 1051–1058.

Li, K.H. (1988). Imputation using Markov chains. *Journal of Statistical Computation and Simulation* **30**, 57–79.

Little, R.J.A. and Rubin, D.B. (1983). On jointly estimating parameters and missing data. *The American Statistician* **37**, 218–220.

Little, R.J.A. and Rubin, D.B. (1987). *Statistical Analysis with Missing Data*. New York: Wiley.

Liu, C. and Liu, J. (1993). Comment on Markov chain Monte Carlo. *Journal of the Royal Statistical Society B* **55**, 82–83.

Liu, C. and Rubin, D.B. (1994). The ECME algorithm: A simple extension of EM and ECM with faster monotone convergence. *Biometrika* **81**, 633–648.

Liu, C. and Rubin, D.B. (1995). ML estimation of the t distribution using EM and its extensions, ECM and ECME. *Statistica Sinica* **5**, 19–39.

Liu, J.S. (1994). The collapsed Gibbs sampler in Bayesian computations with applications to a gene regulation problem. *Journal of the American Statistical Association* **89**, 958–966.

Liu, J.S., Wong, W.H., and Kong, A. (1994). Covariance structure of the Gibbs sampler with applications to the comparisons of estimators and augmentation schemes. *Biometrika* **81**, 27–40.

Liu, J., Wong, W.H., and Kong, A. (1995). Covariance structure and convergence rate of the Gibbs sampler with various scans. *Journal of the Royal Statistical Society B* **57**, 157–169.

Louis, T.A. (1982). Finding observed information using the EM algorithm. *Journal of the Royal Statistical Society B* **44**, 98–130.

Marske, D. (1967). Biochemical oxygen demand data interpretation using sums of squares surface. M.Sc. Thesis, University of Wisconsin.

Matthews, P. (1991). A slowly mixing Markov chain with implications for Gibbs sampling. Technical Report, Department of Mathematics and Statistics, University of Maryland.

McCullagh, P. and Nelder, J. (1989). *Generalized Linear Models*. London: Chapman and Hall.

Mehta, C. and Patel, N. (1983). A network algorithm for performing Fisher's exact test in $r \times c$ contingency tables. *Journal of the American Statistical Association* **78**, 427–434.

Mehta, C. and Patel, N. (1986a). A hybrid algorithm for Fisher's exact test on unordered $r \times c$ contingency tables. *Communications in Statistics* **15**, 387–403.

Mehta, C. and Patel, N. (1986b). FEXACT: a Fortran subroutine for Fisher's exact test on unordered $r \times c$ contingency tables. *ACM Transactions on Mathematical Software* **12**, 295–302.

Meilijson, I. (1989). A fast improvement to the EM algorithm on its own terms. *Journal of the Royal Statistical Society B* **51**, 127–138.

Mendenhall, W.M., Parsons, J.T., Stringer, S.P., Cassissi, N.J., and Million, R.R. (1989). T2 oral tongue carcinoma treated with radiotherapy: Analysis of local control and complications. *Radiotherapy and Oncology* **16**, 275–282.

Meng, X.L. (1994a). Multiple imputation inferences with uncongenial sources of input. *Statistical Science* **9**, 538–573.

Meng, X.L. (1994b). On the rate of convergence of the ECM algorithm. *Annals of Statistics* **22**, 326–339.

Meng, X.L. and Rubin, D.B. (1991). Using EM to obtain asymptotic variance-covariance matrices. *Journal of the American Statistical Association* **86**, 899–909.

Meng, X.L. and Rubin, D.B. (1993). Maximum likelihood estimation via the ECM algorithm: A general framework. *Biometrika* **80**, 267–278.

Metropolis, N., Rosenbluth, A.W., Rosenbluth, M.N., Teller, A.H., and Teller, E. (1953). Equations of state calculations by fast computing machines. *Journal of Chemical Physics* **21**, 1087–1091.

Miller, R. (1980). *Survival Analysis*. New York: Wiley.

Miller, R.G. and Halpern, J.W. (1982). Regression with censored data. *Biometrika* **69**, 521–531.

Morris, C. (1987). Comment on "The calculation of posterior distributions by data augmentation". *Journal of the American Statistical Association* **82**, 542–543.

Mosteller, F. and Wallace, D.L. (1964). *Inference and Disputed Authorship: The Federalist*. Reading: Addison-Wesley.

Müller, P. (1993). A generic approach to posterior integration and Gibbs sampling. Technical Report. Purdue University.

Naylor, J.C. and Smith, A.F.M. (1982). Applications of a method for the efficient computation of posterior distributions. *Applied Statistics* **31**, 214–225.

Odell, P.L. and Feiveson, A.H. (1966). A numerical procedure to generate a sample covariance matrix. *Journal of the American Statistical Association* **61**, 199–203.

Orchard, T. and Woodbury, M.A. (1972). A missing information principle: Theory and applications. *Proceedings of the 6th Berkeley Symposium on Mathematical Statistics*, Vol. 1, pp. 697–715.

Patefield, W.M. (1981). An efficient method of generating $r \times c$ tables with given row and column totals (Algorithm AS 159). *Applied Statistics* **30**, 91–97.

Patil, V.H. (1965). Approximations to the Behrens-Fisher distribution. *Biometrika* **52**, 267–271.

Press, S.J. (1989). *Bayesian Statistics: Principles, Models and Applications*. New York: Wiley.

Raftery, A.E. and Lewis, S.M. (1992). How many iterations in the Gibbs sampler? In: J.M. Bernardo, J.O. Berger, A.P. Dawid and A.F.M. Smith (eds.), *Bayesian Statistics 4*. Oxford: Oxford University Press.

Ramgopal, P., Laud, P.W., and Smith, A.F.M. (1993). Nonparametric Bayesian bioassay with prior constraints on the shape of the potency curve. *Biometrika* **80**, 489–498.

Rao, C.R. (1973). *Linear Statistical Inference and Applications*. New York: Wiley.

Ripley, B. (1987). *Stochastic Simulation*. New York: Wiley.

Ritter, C. and Tanner, M.A. (1992). The Gibbs stopper and the griddy Gibbs sampler. *Journal of the American Statistical Association* **87**, 861–868.

Roberts, G.O. (1992). Convergence diagnostics of the Gibbs sampler. In: J.M. Bernardo, J.O. Berger, A.P. Dawid and A.F.M. Smith (eds.), *Bayesian Statistics 4*. Oxford: Oxford University Press.

Roberts, G.O. and Polson, N.G. (1994). On the geometric convergence of the Gibbs sampler. *Journal of the Royal Statistical Society B* **56**, 377–384.

Robinson, G.K. (1976). Properties of Student's t and of the Behrens-Fisher solution to the two mean problem. *Annals of Statistics* **4**, 963–971.

Roeder, K. (1990). Density estimation with confidence sets exemplified by superclusters and voids in galaxies. *Journal of the American Statistical Association* **85**, 617–624.

Rosenthal, J.S. (1993). Rates of convergence for data augmentation on finite sample spaces. *Annals of Applied Probability* **3**, 319–339.

Rosenthal, J. (1995). Rates of convergence for Gibbs sampling for variance component models. *Annals of Statistics* **23**, 740–761.

Rubenstein, R. (1981). *Simulation and the Monte Carlo Method*. New York: Wiley.

Rubin, D.B. (1987a), Comment on "The calculation of posterior distributions by data augmentation", by M.A. Tanner and W.H. Wong. *Journal of the American Statistical Association* **82**, 543–546.

Rubin, D.B. (1987b). *Multiple Imputation for Nonresponse in Surveys*. New York: Wiley.

Scheffe, H. (1970). Practical solutions of the Behrens-Fisher problem. *Journal of the American Statistical Association* **65**, 1501–1508.

Schenker, N. and Welsh, A.H. (1988). Asymptotic results for multiple imputation. *Annals of Statistics* **16**, 1550–1566.

Schervish, M.J. and Carlin, B.P. (1992). On the convergence of successive substitution sampling. *Journal of Computational and Graphical Statistics* **1**, 111–127.

Schmee, J. and Hahn, G.J. (1979). A simple method for regression analysis with censored data. *Technometrics* **21**, 417–432.

Searle, S.R., Casella, G., and McCulloch, C.E. (1992). *Variance Components*. New York: Wiley.

Sinha, D. (1993). Semiparametric Bayesian analysis of multiple event time data. *Journal of the American Statistical Association* **88**, 979–983.

Sinha, D., Tanner, M.A. and Hall, W.J. (1994). Maximization of the marginal likelihood of grouped survival data. *Biometrika* **81**, 53–60.

Skovgaard, I.M. (1987). Saddlepoint expansions for conditional distributions. *Journal of Applied Probability* **24**, 875–887.

Smith, A.F.M. and Gelfand, A.E. (1992). Bayesian statistics without tears. *American Statistician* **46**, 84–88.

Smith, A.F.M. and Roberts, G.O. (1993). Bayesian computation via the Gibbs sampler and related Markov chain Monte Carlo methods. *Journal of the Royal Statistical Society B* **55**, 3–24.

Smith, A.F.M., Skene, A.M., Shaw, J.E.H., Naylor, J.C., and Dransfield, M. (1985). The implementation of the Bayesian paradigm. *Communications in Statistics* **14**, 1079–1102.

Smith, C.A.B. (1977). Comment on "Maximum likelihood from incomplete data via the EM algorithm", by A.P. Dempster, N.M. Laird and D.B. Rubin. *Journal of the Royal Statistical Society B* **39**, 24–25.

Snell, J.L. (1988). *Introduction to Probability*. New York: Random House.

Spiegelhalter, D., Thomas, A., Best, N., and Gilks, W. (1995a). *BUGS: Bayesian Inference Using Gibbs Sampling*. Cambridge: MRC Biostatistics Unit. (available via anonymous ftp from ftp.mrc-bsu.cam.ac.uk in the directory pub/methodology/bugs)

Spiegelhalter, D., Thomas, A., Best, N., and Gilks, W. (1995b). *BUGS: Examples*. Cambridge: MRC Biostatistics Unit.

Tanner, M.A. (1991). *Tools for Statistical Inference*. New York: Springer-Verlag.

Tanner, M.A. and Wong, W.H. (1987). The calculation of posterior distributions by data augmentation. *Journal of the American Statistical Association* **82**, 528–540.

Tarone, R.E. (1982). The use of historical control information in testing for a trend in proportions. *Biometrics* **38**, 215–220.

Thompson, E.A. and Shaw, R.G. (1990). Pedigree analysis for quantitative traits. *Biometrics* **46**, 399–413.

Thompson, P.A. and Miller, R.B. (1986). Sampling the future: A Bayesian approach to forecasting. *Journal of Business and Economic Statistics* **4**, 427–436.

Tiao, G. and Fienberg, S. (1969). Bayesian estimation of latent roots and vectors with special reference to the bivariate normal distribution. *Biometrika* **56**, 97–104.

Tierney, L. (1994). Markov chains for exploring posterior distributions (with discussion). *Annals of Statistics* **22**, 1701–1762.

Tierney, L. and Kadane, J.B. (1986). Accurate approximations for posterior moments and marginal densities. *Journal of the American Statistical Association* **81**, 82–86.

Tierney, L., Kass, R.E., and Kadane, J.B. (1989). Approximate marginal densities of nonlinear functions. *Biometrika* **76**, 425–434.

Tritchler, D. (1984). An algorithm for exact logistic regression. *Journal of the American Statistical Association* **79**, 709–711.

Tsutakawa, R.M. (1984). Estimation of two-parameter logistic item response curves. *Journal of Educational Statistics* **9**, 263–276.

Tsutakawa, R.M. and Lin, H.Y. (1986). Bayesian estimation of item response curves. *Psychometrika* **51**, 251–267.

Turnbull, B.W., Brown, B.W., and Hu, M. (1974). Survivorship analysis of heart transplant data. *Journal of the American Statistical Association* **69**, 74–80.

Vattulainen, I., Ala-Nissila, T., and Kankaala, K. (1994). Physical tests for random numbers in simulations. *Physical Review Letters* **73**, 2513–2516.

Wakefield, J.C., Smith, A.F.M., Racine-Poon, A., and Gelfand, A.E. (1994). Bayesian analysis of linear and nonlinear population models by using the Gibbs sampler. *Applied Statistics* **43**, 201–221.

Wallace, D.L. (1980). The Behrens-Fisher and Fieller-Creasy problems. In: *R.A. Fisher: An Appreciation*. S. Fienberg and D.V. Hinkley (eds.). New York: Springer-Verlag.

Wang, Y. (1971). Probabilities of the type I errors of the Welch test. *Journal of the American Statistical Association* **66**, 605–608.

Wasserman, L.A. (1989). A robust Bayesian interpretation of likelihood regions. *Annals of Statistics* **17**, 1387–1393.

Wei, G.C.G. and Tanner, M.A. (1990a). A Monte Carlo implementation of the EM algorithm and the poor man's data augmentation algorithm. *Journal of the American Statistical Association* **85**, 699–704.

Wei, G.C.G. and Tanner, M.A. (1990b). Calculating the content and boundary of the HPD region via data augmentation. *Biometrika* **77**, 649–652.

Wei, G.C.G. and Tanner, M.A. (1990c). Posterior computations for censored regression data. *Journal of the American Statistical Association* **85**, 829–839.

Welch, B. (1949). Further notes on Mrs. Aspin's tables. *Biometrika* **36**, 243–246.

Welch, B.L. and Peers, H.W. (1963). On formulae for confidence points based on integrals of weighted likelihoods. *Journal of the Royal Statistical Society B* **25**, 318–329.

Wong, W.H. and Li, B. (1992). Laplace expansion for posterior densities of nonlinear function of parameters, *Biometrika* **79**, 393–398.

Wu, C.F.J. (1983). On the convergence properties of the EM algorithm. *Annals of Statistics* **11**, 95–103.

Zabell, S. (1992). R.A. Fisher and the fiducial argument. *Statistical Science* **7**, 369–387.

Zeger, S. and Karim, R. (1991). Generalized linear models with random effects. *Journal of the American Statistical Association* **86**, 79–86.

Zellner, A. and Rossi, P.E. (1984). Bayesian analysis dichotomous quantal response models. *Journal of Econometrics* **25**, 365–393.

Subject Index

Springer Series in Statistics

(continued from p. ii)

Printed in the United States
26218LVS00003B/121-129

9 780387 946887